Developing Capacities for Teaching Responsible Science in the MENA Region

Refashioning Scientific Dialogue

Committee on Developing a Framework for an International Faculty Development Project on Education About Research in the Life Sciences with Dual Use Potential

Board on Life Sciences
Division on Earth and Life Studies

NATIONAL RESEARCH COUNCIL
OF THE NATIONAL ACADEMIES

In cooperation with

Bibliotheca Alexandrina
and
The World Academy of Sciences (TWAS)

THE NATIONAL ACADEMIES PRESS
Washington, D.C.
www.nap.edu

THE NATIONAL ACADEMIES PRESS 500 Fifth Street, N.W. Washington, DC 20001

NOTICE: The project that is the subject of this report was approved by the Governing Board of the National Research Council, whose members are drawn from the councils of the National Academy of Sciences, the National Academy of Engineering, and the Institute of Medicine. The members of the committee responsible for the report were chosen for their special competences and with regard for appropriate balance.

This study was supported by Grant No. S-LMAQM-10-GR-087 between the National Academy of Sciences and the U.S. Department of State. Any opinions, findings, conclusions, or recommendations expressed in this publication are those of the authors and do not necessarily reflect the views of the organizations or agencies that provided support for the project.

International Standard Book Number-13: 978-0-309-28639-8
International Standard Book Number-10: 0-309-28639-5

Additional copies of this report are available for sale from the National Academies Press, 500 Fifth Street NW, Keck 360, Washington, DC 20001; (800) 624-6242 or (202) 334-3313; *www.nap.edu/*.

Printed in the United States of America.

THE NATIONAL ACADEMIES

Advisers to the Nation on Science, Engineering, and Medicine

The **National Academy of Sciences** is a private, nonprofit, self-perpetuating society of distinguished scholars engaged in scientific and engineering research, dedicated to the furtherance of science and technology and to their use for the general welfare. Upon the authority of the charter granted to it by the Congress in 1863, the Academy has a mandate that requires it to advise the federal government on scientific and technical matters. Dr. Ralph J. Cicerone is president of the National Academy of Sciences.

The **National Academy of Engineering** was established in 1964, under the charter of the National Academy of Sciences, as a parallel organization of outstanding engineers. It is autonomous in its administration and in the selection of its members, sharing with the National Academy of Sciences the responsibility for advising the federal government. The National Academy of Engineering also sponsors engineering programs aimed at meeting national needs, encourages education and research, and recognizes the superior achievements of engineers. Dr. Charles M. Vest is president of the National Academy of Engineering.

The **Institute of Medicine** was established in 1970 by the National Academy of Sciences to secure the services of eminent members of appropriate professions in the examination of policy matters pertaining to the health of the public. The Institute acts under the responsibility given to the National Academy of Sciences by its congressional charter to be an adviser to the federal government and, upon its own initiative, to identify issues of medical care, research, and education. Dr. Harvey V. Fineberg is president of the Institute of Medicine.

The **National Research Council** was organized by the National Academy of Sciences in 1916 to associate the broad community of science and technology with the Academy's purposes of furthering knowledge and advising the federal government. Functioning in accordance with general policies determined by the Academy, the Council has become the principal operating agency of both the National Academy of Sciences and the National Academy of Engineering in providing services to the government, the public, and the scientific and engineering communities. The Council is administered jointly by both Academies and the Institute of Medicine. Dr. Ralph J. Cicerone and Dr. Charles M. Vest are chair and vice chair, respectively, of the National Research Council.

www.national-academies.org

COMMITTEE ON DEVELOPING A FRAMEWORK FOR AN INTERNATIONAL FACULTY DEVELOPMENT PROJECT ON EDUCATION ABOUT RESEARCH IN THE LIFE SCIENCES WITH DUAL USE POTENTIAL

RITA R. COLWELL (*Chair*), Distinguished University Professor, University of Maryland & Johns Hopkins University; Honorary Chairperson and Senior Advisor, Canon U.S. Life Sciences, Inc.
ENRIQUETA C. BOND, President Emeritus, Burroughs Wellcome Fund
JOHN D. CLEMENTS, Professor and Chair, Department of Microbiology and Immunology, School of Medicine; Director, Center for Infectious Diseases, Tulane University
NANCY D. CONNELL, Professor of Medicine, University of Medicine and Dentistry of New Jersey (UMDNJ); Director, UMDNJ Center for Biodefense
CLARISSA DIRKS, Associate Professor of Biology, The Evergreen State College
MOHAMED El-FAHAM, Director, Center for Special Studies and Programs (CSSP), Bibliotheca Alexandrina, Egypt
ALASTAIR W.M. HAY, Professor of Environmental Toxicology and Epidemiology, University of Leeds, United Kingdom
ELIZABETH HEITMAN, Associate Professor of Medical Ethics, Center for Biomedical Ethics, Vanderbilt University Medical Center
ADEL A.F. MAHMOUD, Professor, Woodrow Wilson School, Department of Molecular Biology, Princeton University
MONA MOSTAFA MOHAMED, Professor of Cell Biology, Department of Zoology, Faculty of Science, Cairo University, Egypt
JAMES H. STITH, Vice President Emeritus, Physics Resources Center, American Institute of Physics

STAFF

LIDA ANESTIDOU, Study Director and Senior Program Officer, Institute for Laboratory Animal Research
JO L. HUSBANDS, Scholar and Senior Project Director, Board on Life Sciences
JAY B. LABOV, Senior Scientist, Board on Life Sciences
SAYYEDA AYESHA AHMED, Senior Program Assistant
CARL-GUSTAV ANDERSON, Program Associate, Board on Life Sciences
FRANCES E. SHARPLES, Director, Board on Life Sciences; Acting Director, Institute for Laboratory Animal Research

BOARD ON LIFE SCIENCES

JO HANDELSMAN (*Chair*), Yale University, New Haven, Connecticut
VICKI L. CHANDLER, Gordon and Betty Moore Foundation, Palo Alto, California
SEAN EDDY, HHMI Janelia Farm Research Campus, Ashburn, Virginia
SARAH C.R. ELGIN, Washington University, St. Louis, Missouri
DAVID R. FRANZ, Former Cdr USAMRIID, Consultant, Frederick, Maryland
LOUIS J. GROSS, University of Tennessee, Knoxville
RICHARD A. JOHNSON, Arnold & Porter, LLC, Washington, D.C.
JUDITH KIMBLE, University of Wisconsin, Madison
CATO T. LAURENCIN, University of Connecticut Health Center, Farmington, Connecticut
ALAN I. LESHNER, American Association for the Advancement of Science, Washington, D.C.
BERNARD LO, University of California, San Francisco,
KAREN E. NELSON, J. Craig Venter Institute, Rockville, Maryland
ROBERT M. NEREM, Georgia Institute of Technology, Atlanta
CAMILLE PARMESAN, University of Texas, Austin
ALISON G. POWER, Cornell University, Ithaca, New York
MARGARET RILEY, University of Massachusetts, Amherst
BRUCE W. STILLMAN, Cold Spring Harbor Laboratory, New York
JANIS C. WEEKS, University of Oregon, Eugene
CYNTHIA WOLBERGER, Johns Hopkins University School of Medicine, Baltimore, Maryland
MARY WOOLLEY, Research!America, Alexandria, Virginia

STAFF

FRANCES E. SHARPLES, Director
JO L. HUSBANDS, Scholar/Senior Project Director
JAY B. LABOV, Senior Scientist/Program Director for Biology Education
KATHERINE W. BOWMAN, Senior Program Officer
INDIA HOOK-BARNARD, Senior Program Officer
MARILEE K. SHELTON-DAVENPORT, Senior Program Officer
KEEGAN SAWYER, Program Officer
BETHELHEM M. BANJAW, Financial Associate
CARL G. ANDERSON, Program Associate
SAYYEDA AYESHA AHMED, Senior Program Assistant
ORIN E. LUKE, Senior Program Assistant

Preface

This project merits attention for its success in linking two areas of important work of the U.S. National Academies to advance contributions of the life sciences to health, economic development, and the environment globally. First, the Academies' path-breaking work has focused on how people learn, how effective approaches to teaching can be applied to engage and prepare a new generation of scientists, and how to put that knowledge into practice to transform teaching of undergraduate biology in the United States. Second, influential reports of the National Research Council and the Institute of Medicine published since the late 1980s have helped establish the norms and standards in the United States and internationally for responsible conduct of science. Nevertheless, engagement of the National Academies in responding to concerns that the rapid advances in the life sciences, with their potential for significant benefits, might be misused to cause deliberate harm has provided an important component of this report.

During the last few decades the scientific community has made remarkable progress in developing and promulgating the culture of responsibility that has kept the number of laboratory accidents and cases of deliberate misuse to vanishingly small numbers. But as research capacity extends globally, we need to take advantage of all that is known about how best to instill those standards so that the research enterprise continues to advance knowledge to serve the public and sustain its trust. The project described in this report is, in part, a response to these concerns but it also draws on other work because the Academies' activities are demonstrating that the foundation for effectively engaging the scientific community to address potential risk of misuse is education within a broader framework of the responsible conduct of science. I believe this approach offers the best promise of achieving both security and scientific progress available to all.

Sincere thanks are owed to all members of the committee because their commitment and engagement with the project have been extraordinary. Each worked incredibly hard and several continue to engage with participants in the project to assist in implementing what has been learned in the course of this activity. TheNational Academies staff deserve special recognition, especially Lida Anestidou and Jay Labov, an amazing team. They brought their formidable skills in responsible conduct and scientific teaching, respectively, to the design and implementation of this project. Jo Husbands provided invaluable assistance with planning and oversight of the final report process, and Ayesha Ahmed and Carl-Gustav Anderson were truly remarkable in their research work and administrative support.

Meetings were held in Jordan and Trieste, Italy, and to our hosts in both of those countries a debt of gratitude is acknowledged for their hospitality and professional assistance. The engagement and enthusiasm of participants in the meetings provided encouragement to continue vigorously promoting both scientific teaching and responsible conduct. I wish to acknowledge their essential contributions. I extend my personal thanks, as well as those of the committee, to the Bibliotheca Alexandrina and to The World Academy of Sciences (TWAS), our partners in this project. Their extensive knowledge of the region informed our work and their commitment to scientific excellence made them ideal partners. The task of responsible research in science continues and our hope is that this report will help light the way for global understanding and participation.

—Rita R. Colwell, Chair

Acknowledgments

This report was reviewed in draft form by individuals chosen for their diverse perspectives and technical expertise, in accordance with procedures approved by the National Academies' Report Review Committee. The purpose of this independent review is to provide candid and critical comments that will assist the institution in making its published report as sound as possible and to ensure that the report meets institutional standards for objectivity, evidence, and responsiveness to the study charge. The review comments and draft manuscript remain confidential to protect the integrity of the process.

We thank the following individuals for their review of this report:

Eiman Aleem, University of Arizona and University of Alexandria, Egypt
Rosemary Chalk, Washington, DC
Michael Imperiale, University of Michigan
Peter Mahaffy, The King's University College, Canada
James Revill, University of Sussex, United Kingdom
Henry J. Silverman, University of Maryland
Michelle Withers, West Virginia University

Although these reviewers provided many constructive comments and suggestions, they were not asked to endorse the conclusions or recommendations, nor did they see the final draft of the report before its release. The review of this report was overseen by **Diane Ebert-May**, Michigan State University. Appointed by the National Academies, she was responsible for making certain that an independent examination of the report was carried out in accordance with institutional procedures and that all review comments were carefully considered. Responsibility for the final content of this report rests entirely with the authoring committee and the institution.

Contents

Summary

BACKGROUND

A wave of discoveries in the life sciences, supported by new enabling technologies and drawing on many fields beyond biology, is yielding great social and economic benefits and promises continuing gains in the future. Inspired by this vision, governments across the globe, as well as regional and international organizations, are launching strategies and making investments to apply these advances to address challenges related to food, energy, economic development, the environment, animal and plant health, and human well-being. One of the exciting aspects of this "Century of Biology" is the diffusion of research capacity and infrastructure to many parts of the world, creating an increasingly global life sciences research enterprise.

Along with these hopes and achievements, however, have come concerns about the implications of such rapid advances. Concerns include uneasiness about how an increased understanding of basic life processes, and the resulting potential to manipulate and control them, may result in unintended impacts on the environment or human well-being, or the risk of deliberate misuse of knowledge, tools, and techniques from the life sciences to cause harm.

How the scientific community responds to these concerns can be considered part of the broader relationship between science and society. Beyond its fundamental quest for greater knowledge and understanding, science is conducted in a social context. Science depends on public support, including but not limited to the substantial funding that enables research to take place. Ensuring that scientific research is carried out responsibly is essential to maintaining the relationship between science and society.

The scientific community itself, through its professional bodies and other groups, plays a leading role in fostering and maintaining the norms and standards for what constitutes responsible conduct of science. These standards also provide the basis for training and education about expectations—and in some cases requirements—for professional and responsible behavior. As science becomes an increasingly global enterprise, a growing number of international scientific organizations have joined the activities of national bodies to underscore the ethical imperatives for all involved in scientific research. In addition, a strong tradition of self-governance to maintain responsible conduct in scientific research, often referred to as a "culture of responsibility," provides the foundation for scientists to respond to societal concerns.

Life scientists address ethical and safety issues in their work through three overlapping fields that provide norms and practices to guide research: biosafety, bioethics, and responsible conduct of research. *Biosafety* practices, which have been codified as national and international guidelines, have developed over the last several decades to safeguard the health of laboratory workers and avoid accidental or inadvertent releases of dangerous biological agents and toxins that could harm people or the environment. *Bioethics* encompasses a wide range of ethical issues in different national and disciplinary contexts, including basic research, medical interventions and specifically clinical settings, and protections for human subjects in research. Bioethics also engages many disciplines beyond science and medicine, such as politics, law, philosophy, and theology, so there is great diversity in bioethics education programs. The third field is known by various names, including "research integrity," "scientific integrity," and "research ethics." In the United States, for example, the term "responsible conduct of research" (RCR) emerged in the late 1980s in response to rising concerns about research misconduct. Over time, the mandate evolved into a variably defined set of policies and professional standards that suggested appropriate subjects for instruction.

Where and what material students learn about any of the norms and practices in these fields depends on their area of study, educational institution, and stage of education. They may receive formal instruction ranging from single lectures or online modules to full courses. Informal mechanisms such as mentoring by senior researchers also are important. The scope and quality of education vary widely, but many students still receive little or no exposure to education about responsible conduct of research in the United States, and the problem is worse in other countries. Proposals and initiatives to extend the reach and improve the quality of education for life scientists about responsible conduct of research coincide with and provide a context for a growing interest in education as a fundamental component of efforts to address concerns about deliberate misuse.

THE NATIONAL ACADEMIES'S FACULTY DEVELOPMENT PROJECT

Since the early 2000s, national and international scientific organizations have been engaged in a series of activities to address risks from potential or deliberate misuse of life sciences research. One major line of work has been to inform policymakers about these issues and national and international efforts to minimize, and hopefully prevent, misuse. Another has identified how best to encourage greater engagement by scientists and scientific organizations through education and raising awareness about the importance of responsible conduct in all of its dimensions. The latter activities have set the stage for a major initiative by the National Research Council (NRC) of the U.S. National Academies and its international partners to develop and implement a series of strategic approaches to their education activities. The first part of the initiative applies a model developed by the U.S. National Academies to use active learning methods to improve the quality of undergraduate biology education to the challenges of creating networks of faculty able to teach about dual use issues (see Box 1-1) in the context of responsible conduct of science.[1]

In 2008 the U.S. State Department provided support for an international workshop, convened in Warsaw by several international scientific organizations and organized by the U.S. National Academies and the Polish Academy of Sciences, to:

- survey strategies and resources available internationally for education on dual use issues and identify gaps,
- consider ideas for filling the gaps, including development of new educational materials and implementation of effective teaching methods, and
- discuss approaches for including education on dual use issues in the training of life scientists.

[1] Dual use refers to research that, although undertaken for beneficial purposes, has the potential to yield results that could be misused to cause deliberate harm.

A key feature of the workshop was the inclusion of experts in the growing body of research on the science of learning about how adults learn and what are therefore the most effective approaches to teaching about responsible conduct.

An ad hoc committee under the auspices of the National Academies, with substantial international membership, produced a report from that workshop with a number of conclusions and recommendations for improving education. One of the major recommendations was to create networks of faculty through train-the-trainer programs using active learning approaches drawn from the science of learning (a description of active learning techniques is in Chapter 3 of this report). The networks would provide the basis on which to build sustainable efforts to introduce issues in the context of responsible conduct of science such as dual use. The project described in this report grew out of the recommendations of that workshop.

In 2010, the Biosecurity Engagement Program (BEP) of the U.S. State Department, which provided funding for the Warsaw workshop, agreed to support a two-year project to implement some of the workshop's key recommendations. The full Statement of Task for the project appears in Box S-1. The Middle East–North Africa (MENA) region was chosen to test a prototype that might then be applied in other countries or regions if successful. In addition to the lessons from the Warsaw workshop about the most effective ways to introduce issues of potential misuse it was hoped that combining the best pedagogies with responsible conduct of science would be an appealing capacity-building opportunity for faculty in countries that are interested in using life sciences research for economic growth and improved well-being.

The project was carried out in stages, as shown in the Statement of Task, and overseen by an ad hoc committee of the National Academies, under the auspices of its Board on Life Sciences, with members from the United States, the United Kingdom, and Egypt. It was implemented as a partnership with the Bibliotheca Alexandrina in Alexandria, Egypt, and The World Academy of Science (TWAS), in Trieste, Italy, to draw upon those organizations' extensive ties in the region and increase the chance for the initiative to become sustainable.

The first phase centered on a planning meeting held at TWAS in late spring 2011 to design a general framework for educational institutes for faculty based on the successful model of the National Academies Summer Institute for Undergraduate Biology Education (hereafter NASI) organized by the National Academies and sponsored primarily by the Howard Hughes Medical Institute for undergraduate biology faculty (*www.academiessummerinstitute.org/*). In the project's second phase, the first Institute was held in Aqaba, Jordan, in September 2012 for 28 participants from Algeria, Egypt, Jordan, Libya, and Yemen. It combined sessions devoted to the content of responsible conduct that incorporated various active learning techniques to model what the participants might do in their home institutions. For example, the participants discussed a number of real and hypothetical cases that illustrated different aspects of responsible science, such as authorship and mentorship, the MMR (measles/mumps/rubella) vaccine and autism, and the controversy in 2011 and 2012 over the publication of gain-of-function research related to the H5N1 virus. Additional work in small groups gave them opportunities to use the techniques they were acquiring during the Institute to develop materials that would be useful to their individual academic situations and to present them to other participants prior to returning home.

An online survey shortly after the Institute gathered the participants' initial impressions about their experiences there. In the third and final phase, project participants were invited to apply for small grants to implement some of the combinations of content and methods they designed at the Institute for their home institutions. A small reunion in Amman, Jordan, in April 2013 for the leaders of the teams that received grants enabled the participants to discuss their experiences up to that point, share their insights about the Institute, and consider how their efforts might continue at their institutions and across the MENA region. Their suggestions and lessons provided an important component of the formulation of the committee's findings and conclusions in this report.

INSIGHTS AND REALITIES: LESSONS FROM THE PROJECT

Insights

The NASI model, which involves a variety of evidence-based approaches to active teaching, learning, engagement, and assessment, can be adapted to different topics, cultural contexts, and countries. In the course of reviewing the design and implementation of this Institute, the committee identified a number of insights including logistical, academic, and cultural challenges and realities that could help to improve future projects.

- Active engagement of committee members and Institute leaders before, during, and after the Institute is crucial.
- A detailed application and merit-based selection process can identify enthusiastic and committed participants who will, in turn, demonstrate the importance of such approaches to colleagues at their home institutions and in their disciplines.
- Teaching about and modeling pedagogy can play a significant role in the success of an Institute.
- The demanding pace of the Institute made it hard for some participants to comprehend the concepts and techniques fully and apply them during small group work. Future Institutes will benefit either by providing more time to integrate active learning with new content or by reducing the breadth or both.
- The design of resources and assessments for an Institute benefits from particular attention to linguistic and cultural differences among participants and facilitators. Working with partners from the region where the Institute will take place allows organizers to take into account local customs, traditions, and cultures in ways that remove barriers and foster stronger relationships among organizers and participants.
- The NASI have demonstrated that a reunion of some participants following an Institute can provide new insights about participants' challenges, resources, and opportunities for networking and for sustaining programs (details in Chapter 5). The Institute described in this report further confirmed that a reunion can be especially important for participants from developing countries. For example, by the end of the reunion in Jordan, the scientists who attended agreed that their ability to conduct their own work around responsible conduct and to reach other colleagues at their home institutions, across their individual countries, and in the MENA region as a whole could be expanded and sustained by establishing a network among them. They decided to use this network to share ideas, common challenges,

and opportunities, and to develop joint proposals for future work.

- As with the development of the NASI, new Institutes will require continuing experimentation with and evaluation of all aspects of their design. Feedback from the participants, combined with the results of their projects, can play an important role in future iterations.
- The introduction of both new pedagogies and new content at the same time can be a significant challenge for some participants. Reviewing background materials in advance of the Institute can lessen this impact. However, materials written in English about new concepts, such as active learning and dual use, may present obstacles for non-English speakers.

Realities

- Framing biosafety and dual use issues in the context of responsible science was meaningful to many participants. However, based on conversations during plenary discussions with the participants who attended the reunion meeting in Amman, practical realities such as the lack of basic scientific equipment, reliable Internet connections, and access to scientific journals impede scientists in this region, and especially those from more impoverished nations, from undertaking research at a level where dual use issues raise concerns for them. People undertaking activities where research with dual use potential and/or misuse of technologies is to be one of the topics need to take this reality into account when planning their events or programs.
- Some concepts that are crucial to active learning, responsible science, and dual use cannot be expressed in Arabic. In most of the countries represented at this Institute, teaching about science occurs in English but instructors sometimes provide additional explanations or contexts in Arabic (or French in Algeria). Similarly, Arabic-speaking scientists and students may interpret English words in ways that are different from what the organizers intend. For example, the facilitator team learned that there is only one Arabic word for the two English words "search" and "research," which may contribute to misunderstanding the standards for plagiarism in English-language journals among Arabic-speaking scientists and students. For example, several participants told the group that when they ask their students to define "research," their common response is to find the information in question on Google or another search engine. Hence, these students are not concerned with copying and pasting information from the Internet into their own essays and research reports.
- Scientific research in the MENA region has advanced remarkably over the last generation. Nonetheless, participants reiterated that the lack of a formal framework and infrastructure for research in their countries (e.g., the absence of comprehensive policies and oversight structures regarding authorship, peer review, research with laboratory animals and human subjects, and biosafety) makes it difficult for scientists to follow international standards and to teach best practices in responsible science to their students.
- As the committee learned from the active learning exercise conducted on day 1 of the Institute in which participants from each nation worked together to describe their country's system of higher education (see Chapter 4), there are similarities and differences in education philosophies, approaches to teaching and learning,

facilities, and resources among nations. The differences need to be taken into consideration when planning future Institutes.

- The small grants awarded to participants were used creatively to address an array of educational needs that they identified. In many cases these funds prompted subsequent institutional support to sustain participants' instructional activities. However, as also occurs in the United States, limited funding restricted the ability of these motivated science educators to reach larger audiences who would benefit from instruction on responsible science, biosafety, and dual use issues.
- At the reunion, discussions following each presentation and after all presenters had described their post-Institute activities revealed a great deal of variation in the ways in which participants in those activities were surveyed about their learning and the project's efficacy. Assessment and evaluation are an issue for science faculty across the world. Providing additional guidance and models of survey instruments before such projects are undertaken could provide much more useful and usable data for future initiatives.

Taken together, these insights offer important lessons for the design and implementation of future programs in the MENA region as well as in other parts of the world.

Box S-1
Statement of Task

An ad hoc committee appointed by the National Research Council will develop a framework for an international series of faculty development institutes in key regions around the world with the goal of promoting and enhancing education about issues related to research in the life sciences with dual use potential in the context of responsible conduct of science.

The institutes will bring together higher education faculty in the life sciences as well as experts in related areas to gain greater understanding and experience with methods for effective teaching and learning, develop curricular materials to facilitate education about dual use issues that they will use in their classes, and become prepared to be leaders in their communities on these topics.

The project will be conducted in three phases:

- **Phase I: Planning.** The committee will organize and hold a planning meeting, which will bring together life science educators from the Middle East–North Africa region with leaders in dual use issues and science education. The planning meeting will help to answer substantive and logistical questions that will guide the organization of Phase II, including issues such as scheduling, language, target audience, and evaluation, outreach and dissemination strategies. A consensus letter report will be prepared to guide the organization of Phase II and to serve as a model for organizing similar institutes in the MENA or other regions. In its report, the committee may offer guidance on the distribution of resources to support implementation and follow-up activities.
- **Phase II: First Faculty Development Institute.** The committee will organize a first institute that will feature several invited presentations in addition to workgroups and hands-on exercises. The committee will identify the topics, select and invite speakers and other participants, and work with regional hosts in organizing the session.
- **Phase III: Implementation and Additional Activities.** The committee will work with participants from the first institute to help them implement what they have learned at their home institutions. Small amounts of funding to support implementation, such as the development of new materials, brown bag seminars, or other activities will be made available to at least some of the participating faculty. A follow-up meeting for institute alumni will take be held approximately 6-9 months after the institute, which a small group of staff and committee members will attend.

The committee will also oversee the preparation of a final consensus report that would provide an account of the first institute, the activities initiated by the participants at their home institutions, the discussions at the follow-up meeting of the alumni, and an evaluation of the outcomes. It will also offer further conclusions about successful practices for preparing faculty to teach about research with dual use potential.

Chapter 1
Introduction

THE LIFE SCIENCES AND THE RESPONSIBLE CONDUCT OF RESEARCH

To many in the life sciences community, the 21st century will be the "Century of Biology," just as the 20th was the "Century of Physics" (National Research Council [NRC], 2009a). A wave of discoveries, supported by new enabling technologies and drawing on many fields beyond biology, is yielding great social and economic benefits and holds out the promise of even more widely available gains in the future. Inspired by this vision, national governments as well as regional and international organizations are creating strategies and making investments to apply continuing developments in the life sciences to help solve challenges related to food, energy, economic development, the environment, animal and plant health, and human well-being (see, for example, African Union, 2006; OECD, 2009; Bibliotheca Alexandrina, 2012; White House, 2012).

These accomplishments and ambitions are accompanied, however, by concerns about the implications of such dramatic advances. Concerns include unease about how increased understanding of basic life processes, and the resulting potential to manipulate and control them, may result in unintended impacts on the environment or human well-being as well as the risk of deliberate misuse of knowledge, tools, and techniques from the life sciences to cause harm (NRC, 2002, 2004, 2005; IOM, 2010).

Among a myriad of issues related to the responsible conduct of science, these security issues and the scientific community's response to them can be considered part of the broader relationship between science and society. Beyond its fundamental quest for greater knowledge and understanding, science is conducted in a social context. Science depends on public support, including but not limited to the substantial funding that enables research to take place.

> The ability of science to deliver on its promise of practical and timely solutions to the world's problems does not depend solely on research accomplishments but also on the receptivity of society to the implications of scientific discoveries. That receptivity depends on the public's attitude about what science is finding and on how it perceives the behavior of scientists themselves. (Agre and Leshner, *Science,* 2010:921)

This relationship has important implications for all members of the scientific community.

> Even scientists conducting the most fundamental research need to be aware that their work can ultimately have a great impact on society. Construction of the atomic bomb and the development of recombinant DNA—events that grew out of basic research on the nucleus of the atom and investigations of certain bacterial enzymes, respectively—are two examples of how seemingly arcane areas of science can have tremendous societal consequences. The occurrence and consequences of discoveries in basic research are virtually impossible to foresee. Nevertheless, the scientific community must

> recognize the potential for such discoveries and be prepared to address the questions that they raise. If scientists do find that their discoveries have implications for some important aspect of public affairs, they have a responsibility to call attention to the public issues involved. . . . science and technology have become such integral parts of society that scientists can no longer isolate themselves from societal concerns. (NRC, 1995:20-21)

The relationship between science and society also means that changing social attitudes can affect the conduct of science. The conduct of research in the life sciences has been particularly affected by the continuing evolution of social attitudes and research practices for both human and animal subjects. In response to appalling abuses, standards were created to govern the treatment of human subjects in experiments (Beecher, 1966; The Nuremberg Code, 1949; WMA, 2008; IOM, 2001). The care and use of laboratory animals is another area where standards continue to evolve (NRC, 2011a; IOM, 2011). As the products of science and technology enter the marketplace, both standards and the ethics of practice become critical for environmental safety as well as public health.

The scientific community, through its professional bodies and other groups, plays a leading role in fostering and maintaining the norms and standards for what constitutes responsible conduct of science. As discussed below, these also provide the basis for training and education about the expectations—and in some cases, requirements—for professional and responsible behavior. As science has become an increasingly global enterprise, a growing number of statements and declarations from international scientific organizations have underscored the ethical imperatives for all those involved in scientific research. An early example is the *Declaration on Science and the Use of Scientific Knowledge* from the 1999 World Conference on Science, a collaboration of the International Council for Science (ICSU) and the United Nations Educational, Scientific, and Cultural Organization (UNESCO), which proclaims that

> The practice of scientific research and the use of knowledge from that research should always aim at the welfare of humankind, including the reduction of poverty, be respectful of the dignity and rights of human beings, and of the global environment, and take fully into account our responsibility towards present and future generations,...

and further that

> All scientists should commit themselves to high ethical standards, and a code of ethics based on relevant norms enshrined in international human rights instruments should be established for scientific professions. The social responsibility of scientists requires that they maintain high standards of scientific integrity and quality control, share their knowledge, communicate with the public and educate the younger generation. Political authorities should respect such action by scientists. Science curricula should include science ethics, as well as training in the history and philosophy of science and its cultural impact. (UNESCO, 1999)[2]

The *Singapore Statement,* produced by the Second World Conference on Research Integrity in 2010, includes the principle that "Researchers and research institutions should recognize that they have an ethical obligation to weigh societal benefits against risks inherent in their work" (2nd

---

[2] Key documents from the World Conference on Science are available at *www.unesco.org/science/wcs/*, including the text of the Declaration on Science and the Use of Scientific Knowledge in six languages, *www.unesco.org/science/wcs/eng/declaration_e.htm.*

WCRI, 2010). Similarly, in 2011 the World Science Forum adopted a recommendation on "responsible and ethical conduct of research and innovation."

> In this era of global science, the scientific establishment needs to implement continuous self-reflection to appropriately evaluate its responsibilities, duties and rules of conduct in research and innovation. A universal code of conduct addressing the rights, freedoms and responsibilities of scientific researchers, and the universal rules of scientific research should be shared by the world's scientific community. Furthermore, these rules and policies should be respected by the states and adopted by their national legislations.
>
> Scientists should strengthen their individual and institutional responsibilities to avoid possible harm to society due to ignorance or misjudgment of the consequences of new discoveries and applications of scientific knowledge.
>
> It is the responsibility of those who promote science and scientists to maintain the primacy of moral and social concerns over short-term economic interest in the selection and implementation of industrialised research projects. (World Science Forum, 2011)

In 2012, an international committee convened by the InterAcademy Council (IAC) and IAP—the Global Network of Science Academies (formerly the InterAcademy Panel on International Issues) produced its report on *Responsible Conduct in the Global Research Enterprise*, which among its findings noted that

> Researchers have learned that they cannot dissociate themselves from the uses of the new knowledge they generate. They need to take into consideration the reasonably foreseeable consequences of their own activities. They also have an obligation to participate in the social mechanisms, both within the research community and in the broader society, that explore the implications of research and impose constraints on research if those constraints are justified. (IAC and IAP, 2012:15)

These high-level declarations help set the tone for discussions and can lead to a change in attitudes about the importance of responsible conduct. In 2006, for example, ICSU replaced its Standing Committee on Freedom in the Conduct of Science with a new standing Committee on Freedom *and Responsibility* in the Conduct of Science (emphasis added). While maintaining its traditional strong advocacy for the principles of the universality of science, such as the rights of scientists to travel, associate, and communicate freely, the new committee "differs significantly from its predecessors in that it has been explicitly charged with also considering the responsibilities of scientists" (ICSU, 2008:2).[3] In 2011 the ICSU General Assembly adopted an amendment to the language of its statute on the Universality of Science to recognize formally the importance of responsibility as well as freedom.

> Such practice, in all its aspects, requires freedom of movement, association, expression and communication for scientists, as well as equitable access to data, information, and other resources for research. It requires responsibility at all levels to carry out and communicate scientific work with integrity, respect, fairness, trustworthiness, and transparency, recognising its benefits and possible harms.[4]

---

[3] The ICSU statement on the universality of science may be found at *www.icsu.org/5_abouticsu/INTRO_UnivSci_1.html*.

[4] See *www.icsu.org/about-icsu/structure/committees/freedom-responsibility/statute-5*.

THE "CULTURE OF RESPONSIBILITY" IN THE LIFE SCIENCES

A strong tradition of self-governance to maintain responsible conduct in scientific research, often referred to as a "culture of responsibility" (NRC, 2009b), provides the foundation for scientists to respond to societal concerns. The iconic example of self-governance is the response of the life sciences community in the early 1970s to new gene splicing techniques that would enable them to create recombinant DNA (rDNA) from different organisms. Many of the initial discussions, such as those at a Gordon Research Conference in 1973, concerned potential hazards to laboratory workers or the consequences of an accidental release of rDNA into the environment. This was followed by letters in *Science* and *Nature* from prominent scientists who called for a temporary moratorium on rDNA experiments to enable an assessment of the potential risks. Scientists, as well as some journalists and legal experts, came together in 1975 in the famous Asilomar Conference.[5] The conference concluded that, with appropriate safeguards (i.e., physical and biological containment procedures), most rDNA research could continue. The National Institutes of Health (NIH), which had begun its own reviews of rDNA research in the early 1970s, released *Guidelines for Research Involving rDNA Molecules* in 1976. The guidelines provided procedures and methods for conducting research sponsored by NIH, including a mechanism for reviewing proposed experiments at the institutional level and for adjudicating any cases that could not be resolved there. To extend biosafety procedures to developments in the field of synthetic biology, as of March 2013 the Guidelines were expanded for the first time to cover research "with both recombinant and/or synthetic nucleic acids" (NIH 2012:1).

The activities of scientists and organizations involved in synthetic biology and the response in late 2011 by flu researchers to the controversy over publication of experiments resulting in increased transmissibility of influenza among mammals provide recent examples of voluntary actions.[6] An example of efforts by a government to address potential societal concerns as an integral part of a research program is the Human Genome Project's formal Ethical, Legal, and Social Implications (ELSI) Program (1990–2003).[7]

As discussed in greater detail in Chapter 2, life scientists address ethical and safety issues in their work through three overlapping fields that provide norms and practices to guide research: biosafety, bioethics, and responsible conduct of research. *Biosafety* practices, which are codified in national and international guidelines, have developed over the last several decades to safeguard the health of laboratory workers and avoid accidental or inadvertent releases of dangerous biological agents and toxins that could harm people or the environment.[8] The World Health Organization (WHO) first published its *Laboratory Biosafety Manual* (LBM) in 1983; the third edition came out in 2004 (WHO, 2004). In the United States, the first edition of the *Biosafety in Microbiological*

[5] The Asilomar Conference focused only on the health, safety, and environmental risks of accidentally creating new organisms with dangerous properties.

[6] A group of avian influenza researchers declared a year-long moratorium on further research while international discussions of security and safety issues took place and a number of countries added new measures to address the concerns. A special section in *Science* in May 2012 provides articles from a number of perspectives (*Science*, 2012); the end of the moratorium was announced in January 2013 (Fouchier et al., 2013).

[7] For further information, see *www.ornl.gov/sci/techresources/Human_Genome/project/hgp.shtml*. NIH and the Department of Energy devoted 3-5 percent of their annual project budgets to studying ELSI issues.

[8] For laboratory technicians biosafety training is the primary channel for education about responsible conduct.

and Biomedical Laboratories (BMBL) appeared in 1984; the Centers for Disease Control and Prevention (CDC) and the NIH produced the 5th edition in 2007 (CDC and NIH, 2007). It is important to note that the current editions of both documents have chapters addressing the potential risks of deliberate misuse. In Europe, the 2008 *International Laboratory Biorisk Management Standard* from the European Committee for Standardization (CEN) provided a voluntary management system to support the implementation of specific biosafety practices as well as ways to reduce the risks of misuse (CEN, 2008).

Bioethics is a diverse field and encompasses a wide range of ethical issues in different national and disciplinary contexts, including basic research, medical interventions and specifically clinical settings, and protections for human subjects in research. Bioethics also engages many disciplines beyond science and medicine, such as politics, law, philosophy, and theology, so that there is great diversity in bioethics education programs (see, for example, AAAS, 2008; Revill and Mancini, 2008; Revill, 2009; and Revill et al., 2009).

The third field is known by various names, including "research integrity," "scientific integrity," and "research ethics." In the United States the term "responsible conduct of research" (RCR) emerged in the late 1980s in response to rising concerns about research misconduct. An influential report from the Institute of Medicine (IOM, 1989) recommended systematic education to promote responsible research practices. In 1989 the NIH issued requirements that all those holding certain categories of training grants provide their trainees with instruction in scientific integrity.[9] Over time, the mandate evolved into a variably defined set of policies and professional standards that suggested appropriate subjects for instruction but did not mandate a curriculum or require specific topics. That changed in 2000, when the Department of Health and Human Service's Office of Research Integrity (ORI) issued a policy that required all researchers and research trainees funded by the Public Health Service to undergo training in nine core areas of RCR (ORI, 2000). ORI's policy itself was short lived, but formal programs in RCR instruction continued to grow. Most recently, in November 2009, NIH issued guidelines on topics from which RCR courses could be built. Eight of the subjects are drawn from ORI's original core topics, such as the components of research misconduct (plagiarism, data falsification, and data fabrication) and criteria for authorship, but the new ninth area is "the scientist as a responsible member of society, contemporary ethical issues in biomedical research, and the environmental and societal impacts of scientific research" (NIH, 2009).

Significantly expanding the potential reach of RCR education beyond NIH and biomedical research, in 2009 the National Science Foundation (NSF) mandated that all trainees supported by, or working on, NSF-funded research projects must receive RCR instruction. NSF is the major funder of basic research in the broader life sciences, including fundamental sciences in agriculture, and also supports fields such as physical sciences, engineering, and computer sciences that play growing roles in the increasingly integrated world of life sciences research (NRC, 2010). Given that NIH and NSF fund international scientists and collaborations, their expanded requirements have a global impact. These initiatives by U.S. funders complement a growing international effort to raise awareness of responsible science and promote RCR education, for example through the series of World Conferences on Research Integrity and the concomitant statements on

[9] The requirement was expanded to cover all training grant recipients in 1992 and expanded further in 2009.

various aspects of research integrity issued by them.[10] The first World Conference was held in Portugal in 2007, the second in Singapore in 2010, and the third in Canada in 2013.

Where and what students learn about any of the norms and practices depends on their field of study, institution, and stage of education. They may receive formal instruction ranging from single lectures or online modules to full courses; and informal mechanisms such as mentoring by senior researchers play an essential role. As respected members of the community, mentors serve as important messengers for the norms of the profession.

The scope and quality of available education varies widely, but many students still receive little or no exposure to education about responsible conduct of research. The proposals and initiatives to extend the reach and improve the quality of education for life scientists about responsible conduct of research, such as those described above, coincide with and provide a context for a growing interest in education as a fundamental component of efforts to address concerns about deliberate misuse. The next section discusses this development further.

EDUCATION AS THE FOUNDATION FOR RESPONSIBLE CONDUCT OF RESEARCH

The Life Sciences and the "Web of Prevention"

One of the concerns that has arisen in response to the rapid advances in the life sciences is the potential risk that the knowledge, tools, and techniques resulting from these discoveries might be misused to cause deliberate harm. These concerns come in the wider context of a dramatically changed international security environment, where threats from nonstate actors—including a potential willingness to use weapons of mass destruction (WMD)—are considered as grave as those nation-states could pose (United Nations, 2004). In the United States, for example, the attacks on September 11, 2001, and the anthrax mailings a month later heightened these concerns dramatically and focused attention on harmful uses of biological agents and toxins on a large scale.[11] At the same time, the publication of a number of scientific articles early in the 2000s sparked debates about whether the published methods and results of certain types of experiments could provide a "blueprint" or "roadmap" for those who sought to cause harm.[12]

It is noteworthy, however, that the research that raised the most concern about potential misuse in many cases also promised important potential benefits. Then and now, judgments about relative risks and rewards were seldom simple or definitive (NRC, 2004; *Science,* 2012). The difficulties and uncertainties associated with assessing whether and how the results of life sciences research intended for legitimate and

[10] The 2010 Singapore Statement on Research Integrity is available at *www.singaporestatement.org/statement.html,* and the 2013 Montreal Statement on Research Integrity in Cross-Boundary Research Collaborations at *www.wcri2013.org/Montreal_Statement_e.shtml.*

[11] In October 2001, letters containing anthrax were sent to offices of several media organizations in the United States as well as to members of Congress. Five people eventually died, including postal workers who were exposed to anthrax spores that escaped the letters. An FBI investigation concluded that the letters had been sent by a scientist at the U.S. Army Research Institute for Infectious Diseases (NRC, 2009b).

[12] Some of the key articles are discussed in *Biotechnology Research in an Age of Terrorism* (NRC, 2004:25-29). Epstein (2001) reviews the issues and policy options under discussion at the time; Zilinskas and Tucker (2002) reflect the concerns in the security policy community. These discussions have not abated. For example, many similar concerns were raised more recently about publications related to the sequencing of the influenza virus from the 1918 pandemic (van Aken, 2006; CDC, 2006) and synthetic mutations in the H5N1 virus (*Science,* 2012).

beneficial purposes could be misused is sometimes referred to as the "dual use dilemma" (NRC, 2004:1).[13] That term and a number of others associated with potential misuse remain the subject of considerable confusion and debate. Box 1-1 provides definitions and brief discussions of some of the key terms as they are used in this report.

It is important to underscore that the current concerns extend beyond the infectious disease agents that were the focus of past state-level biological weapons programs (Wheelis et al., 2006). Two examples are advances in neuroscience (Royal Society, 2012) and the promise of constructing living organisms de novo through synthetic biology (Tucker and Zilinskas, 2006; Garfinkel et al., 2007; Mukunda et al., 2009).[14]

Investigators in many areas of the life sciences could be affected even if their particular research poses no apparent risks. Policy actions taken in response to perceptions about a particular field or research focus could have direct but also larger indirect consequences for the research enterprise.[15] A shift in public perceptions to see more risks than rewards from expanding knowledge and capabilities will have repercussions for all life scientists. A number of studies have recommended that life scientists need to become more aware of and engaged in discussions about potential misuse of their work, as well as the positive contributions they can make to crafting and implementing strategies and policies to support continued scientific progress while preventing harm (Royal Society, 2004; NRC, 2004, 2006a, 2011c; IAP, 2005; WHO, 2007; IAC and IAP, 2012). The preferred path to awareness and engagement is generally through widespread education about potential risks and how responses fit within the broader perspective of responsible conduct of science and scientific research. For example, the second phase of the IAC-IAP project that produced *Responsible Conduct in the Global Research Enterprise* (IAP-IAC, 2012) will create educational materials, based in part on the model of the widely adopted handbook, *On Being a Scientist,*[16] from the National Research Council (NRC) of the U.S. National Academies.[17] The IAC-IAP resources are intended to be used by national and regional scientific organizations to promote discussion about what responsible conduct means in practice.

The project described in this report is part of the work of a number of national and international scientific organizations to put such recommendations about engaging scientists into practice. As Chapter 3 discusses, it is also clear from the emerging research literature on human

[13] Efforts to foster attention to dual use issues extend beyond the life sciences and research ethics to include other fields of science, engineering, and health; NRC (2007a) provides an example from the United States.

[14] The implications of these and other developments are discussed in a report prepared by several national and international scientific organizations (NRC, 2011b).

[15] In the United States, for example, the Select Agent Program administers an extensive set of regulations governing approximately 80 biological agents and toxins that affect humans, plants, and animals. For an account of the development and implementation of the program see NRC (2009b); current information is available at www.selectagents.gov/.

[16] The third edition (NRC, 2009c) is available at *www.nap.edu/catalog.php?record_id=12192*; the second edition (1995) at *www.nap.edu/catalog.php?record_id=4917*.

[17] The National Academies is the collective name for four private, nonprofit U.S. institutions: the National Academy of Sciences, the National Academy of Engineering, the Institute of Medicine, and the National Research Council. Further information is available at *www.nas.edu*.

BOX 1-1
Definitions of Key Terms

Dual Use

Traditionally, "dual use" refers to items that have both commercial and military applications. Obvious examples are helicopters and computers, particularly high-performance ones. It may also have positive connotations for the "spin-off" of military research and development to benefit the civilian economy. Research and equipment that supports dual use products may also fall into the dual use category; very broadly, basic research might not usually be considered dual use, whereas applied research would.

Concerns arising in the mid- to late 1990s and early 2000s that the results of research in the life sciences might be misused to cause deliberate harm led to a different use of the term "dual use": research intended for beneficial purposes that could be misused for malevolent purposes (see, for example, NRC, 2004). In an attempt to define what should be the appropriate focus of efforts to prevent misuse, the U.S. National Science Advisory Board for Biosecurity proposed a specialized category called "dual use of concern" (DURC), which it defined in 2007 as "research that, based on current understanding, can be reasonably anticipated to provide knowledge, products, or technologies that could be directly misapplied to pose a threat to public health and safety, agricultural crops and other plants, animals, the environment, or materiel" (NSABB, 2007). More recently, the World Health Organization adopted the term dual use research of concern for an international workshop on oversight of research in the wake of the H5N1 controversy (see WHO, 2013). Its definition of DURC is "life sciences research intended for benefit, but with results which might easily be misapplied to produce harm" (WHO, 2013:1).

learning and cognition that learners are able to understand issues more deeply, acquire knowledge more easily, and retain it for longer periods of time when they actively engage with them rather than confronting them more passively (e.g., by listening to lectures).

The challenge of engaging scientists in helping to mitigate the potential misuse of life sciences is part of what some in the international law and security community have proposed as a "web of prevention" (Rappert and MacLeish, 2007).[18] A central element of this web is the international norm against the use of disease as a weapon, embodied in two agreements: the 1925 Geneva Protocol and the 1972 Biological and Toxin Weapons Convention (BWC).[19] The BWC was the first international treaty to ban an entire class of weapons.[20] BWC States Parties are

[18] The term was coined by the International Committee of the Red Cross in 2002 as part of its "Biotechnology, Weapons, and Humanity" campaign.

[19] The formal title of the Geneva Protocol, which prohibits first use of chemical and biological weapons, is the "Protocol for the Prohibition of the Use in War of Asphyxiating, Poisonous or Other Gases, and of Bacteriological Methods of Warfare." The BWC's formal title is the "Convention on the Prohibition of the Development, Production and Stockpiling of Bacteriological (Biological) and Toxin Weapons and on Their Destruction." These two agreements address threats from nation-states; the 2004 UN Security Council Resolution 1540 extends the prohibitions to cover nonstate actors.

[20] The 1997 Chemical Weapons Convention (CWC) is the second WMD prohibition treaty. The increasing convergence of chemistry and biology in research and applications is also fostering greater connections between the BWC and the CWC.

Biosafety and Biosecurity[a]

Two widely available definitions of these terms are:

Biosafety: "Laboratory biosafety describes the containment principles, technologies and practices that are implemented to prevent the unintentional exposure to pathogens and toxins, or their accidental release" (World Health Organization [WHO], 2006:iii).

Biosecurity: "the protection, control and accountability for valuable biological materials [including information] ... within laboratories in order to prevent their unauthorized access, loss, theft, misuse, diversion or intentional release" (WHO, 2006:iii).

Confusion about the terms raises two different types of issues. The most basic is that in quite a few languages the term "biosecurity" does not exist or is identical with "biosafety." French, Spanish, and other Romance languages, as well as German, Russian, and Chinese illustrate this practical problem.

The more serious problem for biosecurity is that the term is already in widespread use for a number of other international issues. For example, to many "biosecurity" refers to the obligations undertaken by states adhering to the Convention on Biodiversity and particularly the Cartagena Protocol on Biosafety, which is intended to protect biological diversity from the potential risks posed by *living modified organisms* resulting from modern biotechnology.[b] "Biosecurity" has also been narrowly applied to efforts to increase the security of dangerous pathogens, either in the laboratory or in dedicated collections; guidelines from both the World Health Organization (WHO, 2004) and the Organization for Economic Cooperation and Development (OECD, 2007) use this more restricted meaning of the term. In an agricultural context, the term refers to efforts to exclude the introduction of plant or animal pathogens.

[a] This section is taken from NRC, 2011c:20-21.

[b] Further information on the convention is available at *www.cbd.int/convention/* and on the Protocol at *www.cbd.int/biosafety/*.

obligated to "take any necessary measures" appropriate to their legal processes to carry out the treaty's goals. In addition, countries and some regional organizations are increasingly promulgating laws, regulations, and guidelines to address potential misuse directly or to contribute indirectly through the governance of research and commercial activities.

The concept of a web also assigns an essential role to measures of self-governance, including guidelines, "soft law," codes of conduct, and other voluntary practices that could have both nongovernment and possibly government components. Institutions such as universities, nongovernmental organizations, and scientific organizations are providing essential "bottom-up" initiatives (NRC, 2009d; Rappert, 2010). These complement the prospects for "top-down" attention or initiatives on the part of international bodies such as the World Health Organization (WHO, 2005, 2007, 2013) and the Organization for Economic Cooperation and Development (OECD, 2004) and from the activities associated with the operation and implementation of the BWC cited above.[21]

[21] The papers and presentations during the 2008 meetings of experts and states parties and the 7th BWC Review Conference in December 2011 provide a number of examples. They also underscored the need for the States Parties to the Convention to take a more active role in supporting the bottom-up initiatives. For further information, see the "Meetings and Documents" section on the BWC website (*www.unog.ch/80256EE600585943/(httpPages)/92CFF2CB73D4806DC12572BC00319612?OpenDocument.*

Origins of the Project

Since the early 2000s, national and international scientific organizations have been engaged in a series of activities to address the risks of potential misuse of the results of life sciences research to cause deliberate harm in the context of responsible conduct of science. One major line of work has been informing policymakers about the implications of trends in the life sciences for the implementation of national and international efforts to prevent misuse, both in terms of potential risks and the contributions that science and technology can make to reducing them (Royal Society, 2006; NRC, 2011b). Another has been identifying how best to encourage greater engagement by scientists and scientific organizations through education and raising awareness (NRC, 2009d; 2011c).[22] The latter activities set the stage for a major new initiative by the National Academies and its international partners to develop and implement a strategic approach to their education activities.

In 2008 the U.S. Department of State asked the IAP to convene a workshop to:

- survey strategies and resources available internationally for education on dual use issues and identify gaps,
- consider ideas for filling the gaps, including development of new educational materials and implementation of effective teaching methods, and
- discuss approaches for including education on dual use issues in the training of life scientists. (NRC, 2011c:2)

The workshop (hereafter the Warsaw workshop) was organized as a collaboration of IAP with several other international scientific organizations; the Polish Academy of Sciences served as the host in collaboration with an ad hoc committee with substantial international membership under the auspices of the National Academies.

The meeting, which combined plenary sessions to introduce topics and breakout sessions to permit discussions in depth, brought together more than sixty experts from just under thirty countries and several international organizations. The participants included active researchers from a range of fields in the life sciences, specialists in bioethics and biosecurity, and, as one of the workshop's special features, experts in the science of learning. This mix of backgrounds and expertise underscored the two themes at the heart of the workshop's design:

- To engage the life sciences community, the particular security issues related to research with dual use potential would best be approached in the context of responsible conduct of research, the wider array of issues that the community addresses to fulfill its responsibilities to society.
- Education about dual use issues would benefit from the insights of the "science of learning," the growing body of research about how individuals learn at various stages of their lives and careers and the most effective methods for teaching them, which provides the foundation for efforts in many parts of the world to improve the teaching of science and technology at all levels of instruction. (NRC, 2011c:3)

The workshop also discussed the similar challenges faced by any effort to introduce new material, such as the competition for space in an already crowded curriculum, or an academic reward structure that did not put high value on innovation or excellence in teaching. One clear message was "the importance of identifying and

[22] Both of these reports, undertaken with a number of international partners, include accounts of work by national and international scientific organizations.

supporting 'champions' to the success of initiatives" (NRC, 2011c:87). In addition, participants consistently cited the limited number of faculty and instructors able to teach about dual use issues. This led to an extensive discussion of the importance of networks to support and sustain efforts to introduce new topics and new approaches. A number of examples related to dual use that also drew on the research about effective teaching—such as online faculty development courses from the University of Bradford in the United Kingdom and the WHO train-the-trainer courses on biosafety and biosecurity redesigned to escape an older "death by PowerPoint" approach—offered potential models for new efforts.[23]

> For all the approaches participants stressed the importance of including plans for post-training support, both for developing and implementing new methods and materials and for sharing lessons learned and best practices. It is worth noting that some models... deliberately include small teams rather than single individuals from a given institution in order to enhance the chances of sustaining what is learned and a commitment to implementation is part of the selection process. The champions...may also help to create and sustain a more hospitable climate for new content and methods. In addition to supporting work at home institutions, some models for building networks of faculty and instructors also bring graduates together after their training for special follow-up activities to reinforce what was learned, while others rely on the normal cycle of meetings that take place in a discipline or professional field to provide convening opportunities (NRC, 2011c:89).

The discussion also included some models for more general faculty development that could be adapted, in particular the National Academies Summer Institutes for Undergraduate Biology Education (NASI) that became the basis for the project described in this report.[24]

The NRC committee took responsibility for producing the report, which contained a number of conclusions and recommendations. Selected conclusions relevant to this project and the full list of recommendations may be found in Appendix A, but one specific recommendation is particularly relevant.

> Build networks of faculty and instructors through train-the-trainer programs, undertaking this effort if possible in cooperation with scientific unions and professional societies and associations. (NRC 2011c:9-10)

CREATING NETWORKS OF FACULTY: THE MIDDLE EAST–NORTH AFRICA PROJECT

In 2010, the Biosecurity Engagement Program (BEP) of the U.S. Department of State, which provided the funding for the Warsaw workshop, agreed to support a two-year project to implement some of the workshop's key recommendations. The broad goal of the project was to "develop a framework for an international series of faculty development institutes in key regions around the world with the goal of promoting and enhancing education about issues related to research in the life sciences with dual use potential in the context of responsible conduct of science." The full

[23] Two examples of other dual use–related projects that have taken place since the Warsaw workshop that include active learning are EUBARnet (2012) and Novossiolova et al. (2013).

[24] The general characteristics of faculty development programs, one variant of train-the-trainer models, are discussed in Chapter 3.

Statement of Task (SOT) for the project is shown in Box 1-2.

The project was overseen by an ad hoc NRC committee with members from the United States, Europe, and Egypt (see Appendix E). The committee interpreted the "framework" in the SOT as concerned with the design of the institutes and not the development of underlying concepts. The project in fact builds on the concepts related to responsible science and dual use issues developed in the course of almost a decade of work by the National Academies and other organizations already discussed in this chapter, as well as on other concepts related to active learning described in Chapter 3 that reflects a comparably long National Academies engagement.

This report is intended to be useful to a number of audiences:

- Scientists in the Middle East–North Africa (MENA) region and elsewhere who may not have considered the issues addressed in the Institute and want information about the concepts associated with responsible science and ideas about how to introduce the material into their classrooms and institutions.
- Program managers and funders who might support projects related to dual use issues, responsible conduct, or capacity building in the life sciences and be interested in new approaches.
- Experts in responsible conduct who might not be familiar with active learning techniques.
- Experts in active learning who might not have considered how the approaches could be applied to new areas.

It has a strong emphasis on practical implementation and tries to provide sufficient detail to give readers a sense of how similar institutes might be adapted and organized in other contexts.

The BEP program operates in many parts of the world, but it emphasizes certain regions and priority countries with them. After consultation with the sponsor, the Middle East–North Africa (MENA) region was chosen to test a prototype that could then be applied in other countries or regions. In addition to the lessons from the Warsaw workshop about the most effective ways to introduce issues of potential misuse, the committee hoped combining the best pedagogy with responsible conduct of science would be an appealing capacity-building opportunity for faculty in countries interested in using life sciences research for economic growth and improved wellbeing.

The project was carried out in stages as a partnership with the Bibliotheca Alexandrina in Alexandria, Egypt, and The World Academy of Sciences (TWAS), in Trieste, Italy (see Appendix B). The two institutions' standing and extensive networks in the region were essential to the effective implementation of the project. Unfortunately, continuing political uncertainties in the MENA region in the wake of the Arab Spring necessitated a number of delays and changes, prolonging the project by about a year. The first phase centered on a planning meeting held at TWAS in late spring 2011 to design a general framework for educational institutes for faculty based on the NASI model; a description of NASI is provided in Chapter 3.[25] In the project's second phase, discussed in detail in Chapter 4, the first Institute was held in Aqaba, Jordan, in September 2012 for 28 participants from Algeria, Egypt, Jordan, Libya, and Yemen. An online survey shortly after the Institute gathered the participants' initial impressions. In the third and final phase, project participants applied for small grants to implement some of

---

[25] In the context of this report, the terms "workshop" and "institute" are interchangeable.

the content/methods combinations they designed at the Institute in Jordan in their home institutions.[26] In April 2013 a small reunion for the leaders of the teams that received grants in Amman, Jordan, enabled the participants to discuss their experiences up to that point and also share their insights about the Institute. Their suggestions and lessons provided important input into the formulation of the committee's findings and conclusions for this report.

STRUCTURE OF THE REPORT

This chapter has provided an introduction to how concerns about potential misuse of advances in the life sciences can be addressed in the context of responsible conduct of science and the essential role that education plays in fostering the engagement of the scientific community in responses that seek both security and continued scientific progress.

Chapter 2 elaborates on the development and current status of the basic concepts and approaches to education and training in the responsible conduct of science. Chapter 3 provides an overview of the science of learning, research that reveals how people learn and how to use the insights on human learning and cognition to improve teaching practices. As noted above, commitment to supporting the best possible pedagogy is a key feature of the MENA project. These two chapters are intended to offer quick primers for readers with expertise in one but not necessarily both of the subjects. Chapter 4 describes the planning meeting and the first Institute, held in Jordan in September 2012, while Chapter 5 discusses the activities undertaken by participants after the Institute to implement what they learned. Chapter 6 offers a preliminary evaluation of the Institute, along with the committee's findings, conclusions, and ideas for the future.

---

[26] The project was able to support five grants and BEP provided funds to support another three.

Box 1-2
Statement of Task

An ad hoc committee appointed by the National Research Council will develop a framework for an international series of faculty development institutes in key regions around the world with the goal of promoting and enhancing education about issues related to research in the life sciences with dual use potential in the context of responsible conduct of science.

The institutes will bring together higher education faculty in the life sciences as well as experts in related areas to gain greater understanding and experience with methods for effective teaching and learning, develop curricular materials to facilitate education about dual use issues that they will use in their classes, and become prepared to be leaders in their communities on these topics.

The project will be conducted in three phases:

- **Phase I: Planning.** The committee will organize and hold a planning meeting, which will bring together life science educators from the Middle East–North Africa region with leaders in dual use issues and science education. The planning meeting will help to answer substantive and logistical questions that will guide the organization of Phase II, including issues such as scheduling, language, target audience, and evaluation, outreach and dissemination strategies. A consensus letter report will be prepared to guide the organization of Phase II and to serve as a model for organizing similar institutes in the MENA or other regions. In its report, the committee may offer guidance on the distribution of resources to support implementation and follow-up activities.
- **Phase II: First Faculty Development Institute.** The committee will organize a first institute that will feature several invited presentations in addition to workgroups and hands-on exercises. The committee will identify the topics, select and invite speakers and other participants, and work with regional hosts in organizing the session.
- **Phase III: Implementation and Additional Activities.** The committee will work with participants from the first institute to help them implement what they have learned at their home institutions. Small amounts of funding to support implementation, such as the development of new materials, brown bag seminars, or other activities will be made available to at least some of the participating faculty. A follow-up meeting for institute alumni will take be held approximately 6-9 months after the institute, which a small group of staff and committee members will attend.

The committee will also oversee the preparation of a final consensus report that would provide an account of the first institute, the activities initiated by the participants at their home institutions, the discussions at the follow-up meeting of the alumni, and an evaluation of the outcomes. It will also offer further conclusions about successful practices for preparing faculty to teach about research with dual use potential.

Chapter 2
Responsible Conduct and Integrity in Science

This chapter is intended to provide additional information to build on the discussion in Chapter 1 of the development of concepts of scientific responsibility and the ways in which those concepts are taught to students and practitioners. The integrity of research and ethical grounding of science have been prominent concerns of international research institutions, professional societies, and funding agencies for the past three decades. In the 1980s, professional and governmental attention to the growth and increasing complexity of life sciences research led to multidisciplinary efforts to define research integrity and responsible conduct of research in concrete terms. Much of this effort started in the United States, where it gained prominence in response to high-profile cases of research misconduct—the fabrication or falsification of research data and the theft of others' ideas, words, and data through plagiarism. Since those early years, a large and increasingly comprehensive body of standards for ethical and scientifically sound research practices has developed, and researchers at many levels are encouraged to pursue formal study and dissemination of these practices. Ongoing instruction in responsible conduct of research is now commonly accepted in science education, particularly in pre- and postdoctoral training. The integrity of the research process is recognized to be "critical for excellence, as well as public trust, in science" (NSF, 2009).

EVOLVING TERMINOLOGY AND DEFINITIONS

As with most new concepts, key terminology related to research integrity has evolved with discussion of its central themes and specific issues. The larger concept of integrity in science was first formally defined in two reports from the National Academies (IOM, 1989; NRC, 1992). Both used the term integrity in a way that emphasized researchers' *honesty.* In 1989, the Institute of Medicine's Committee on the Responsible Conduct of Research in the Health Sciences defined *integrity in research* to mean "that the reported results are honest and accurate and are in keeping with generally accepted research practices" (IOM, 1989:v). In 1992, the Panel on Scientific Responsibility and the Conduct of Research published the report *Responsible Science*, which highlighted the *integrity of the research process.* In this context, the committee defined integrity as "*the adherence by scientists and their institutions to honest and verifiable methods in proposing, performing, evaluating, and reporting research activities*" (NRC, 1992:17).

While honesty remains the focal point of research integrity, today's definitions typically include regulatory compliance and adherence to professional standards in the research process. For example, the U.S. National Institutes of Health policy guide defines research integrity as:

- The use of honest and verifiable methods in proposing, performing, and evaluating research;
- Reporting research results with particular attention to adherence to rules, regulations, guidelines; and
- Following commonly accepted professional codes or norms. (NIH, 2012)

EDUCATION IN THE RESPONSIBLE CONDUCT OF RESEARCH

The term *responsible conduct of research*—frequently referred to by its acronym RCR—emerged during this same period as research funders and academic research institutions endeavored to distinguish research misconduct from the processes and activities that constituted good scientific practice (IOM, 1989; NIH, 1990; NRC, 1992). The concept became particularly important in education policy following a 1990 amendment to the NIH's policies on research training grants. The amendment required the mandatory instruction in RCR that was part of all institutional research training grants to add instruction on professional ethics and regulatory standards (NIH, 1990).

Initially the content of such instruction was not defined. Formal textbooks and other curricular materials developed both before and in response to the training grant mandate covered a wide array of issues that grew both broader and more concrete over time (Heitman and Bulger, 2005). For example, the first edition of the National Academies' *On Being a Scientist*, published in 1989, examined the nature of scientific research and the social mechanisms of science from a largely historical and sociological perspective (NRC, 1989). In 1995, the second edition, subtitled *Responsible Conduct in Research*, expanded its discussion of the social and historical context of science to incorporate more instruction on professional standards of practice and the scientist's role in society (NRC, 1995).

Over the past two decades, however, instruction in responsible conduct of research has increasingly focused on the elements of research practice and the ethical values and professional norms of science. Current NIH policy on research training grants defines responsible conduct of research as "the practice of scientific investigation with integrity," which includes "awareness and application of established professional norms and ethical principles in the performance of all activities related to scientific research" (NIH, 2009).

In 2000, a decade after NIH's initial training grant mandate for instruction in responsible conduct of research, the Department of Health and Human Services' Office of Research Integrity (ORI) proposed a new educational policy to extend NIH's requirement for RCR instruction in training grants to *everyone* funded by Public Health Service grants, not just research trainees (ORI, 2000). This policy was short lived, due largely to the anticipated costs of providing such an extensive educational activity across the federally funded research enterprise (Steneck and Bulger, 2007). Nonetheless, the policy's impact on education was significant in that ORI defined nine core areas for instruction that contained the knowledge, skills, and attitudes essential to responsible conduct. These nine core areas were:

- Data acquisition, management, sharing, and ownership
- Mentor/trainee responsibilities
- Publication practices and responsible authorship
- Peer review
- Collaborative science
- Human subjects
- Research involving animals

- Research misconduct
- Conflict of interest and commitment.[27]

Although ORI suspended the policy in 2001, its nine core instructional areas provided a common framework for the development of practice standards and research policy, as well as a wide range of educational resources. Even after the policy was withdrawn, various curricular materials, including the National Academies' third edition of *On Being a Scientist* (NRC, 2009c), were revised and expanded to address the nine core areas and explore case studies in which relevant professional standards were at issue.

In 2009, both NIH and the U.S. National Science Foundation updated their requirements for instruction in responsible conduct of research, enlarging, reconfiguring, and reprioritizing the core areas and including, for the first time, formal attention to the practices related to biosafety and research with dual use potential (NIH, 2009; NSF, 2009). Today's research integrity educators are now called upon to emphasize the following core areas of responsible conduct:

- Conflict of interest
- Policies regarding human subjects, live vertebrate animal subjects in research, and safe laboratory practices
- Mentor/mentee relationships and responsibilities
- Collaborative research, including collaborations with industry
- Peer review
- Data acquisition and laboratory tools, management, sharing and ownership
- Research misconduct
- Responsible authorship and publication practices and
- The scientist as a responsible member of society, contemporary ethical issues in biomedical research, and the environmental and societal impacts of scientific research.[28]

INTERNATIONAL PERSPECTIVES ON RESEARCH INTEGRITY AND RESPONSIBLE CONDUCT OF RESEARCH

Since the early 2000s, several multinational professional organizations have worked to elaborate practice standards and ethical norms for worldwide adoption, particularly in the life sciences. The European Science Foundation (ESF), a multinational organization with member societies in 23 countries, issued its first major statement on research integrity, *Good Scientific Practice in Research and Scholarship*, in December 2000 (ESF, 2000). At that time, a variety of member organizations had developed country-specific policies on research misconduct and guidelines on responsible research, but these standards were not well integrated. ESF's statement emphasized the importance of professional governance and researchers' honesty at all stages of scientific inquiry. ESF called for member organizations to develop both national and European-level codes of good scientific practice and to pursue the harmonization of national standards.

In December 2007, ESF and ORI published *Research Integrity: Global Responsibility to Foster Common Standards* (ESF, 2007), a catalogue of international activities in research integrity that also reported on the workshop Best Practices for Ensuring Scientific Integrity and Preventing Misconduct, sponsored by the Organization for

[27] See http://oprs.usc.edu/files/2013/01/PHS_Policy_on_RCR1.pdf.

[28] See http://oprs.usc.edu/files/2013/01/PHS_Policy_on_RCR1.pdf.

Economic Cooperation and Development's (OECD) Global Science Forum, which had sought to foster international cooperation in the development of policy and administrative systems in international science (OECD, 2007). A related expert group from the European Commission (EC) recommended that the Commission take the lead in developing European standards, harmonizing definitions and principles, and investigate emerging issues in transnational research. The following year, ESF published *Stewards of Integrity* (ESF, 2008), a review of European policies and programs that supported good scientific practices. ESF's survey found governmental and nongovernmental organizations in all of its member countries that had begun to articulate standards of responsible research.

In 2010, ESF issued a background report, *Fostering Research Integrity in Europe*, which outlined a framework for shared governance of research integrity and recommended that ESF and All European Academies (ALLEA) endorse European standards (ESF, 2010). In 2011, the ESF Member Organization Forum on Research Integrity and ALLEA finalized the European Code of Conduct for Research Integrity (ESF and ALLEA, 2011). The Code identified eight principles for all researchers, research organizations, universities and funders to observe:

- Honesty in communication
- Reliability in performing research
- Objectivity
- Impartiality and independence
- Openness and accessibility
- Duty of care
- Fairness in providing references and giving credit
- Responsibility for the scientists and researchers of the future. (ESF and ALLEA, 2011:5)

On a wider, global level, IAP–The Global Network of Science Academies (formerly the InterAcademy Panel on International Issues), an organization of over 100 national academies of science, has also been a leader in promoting research integrity and responsible conduct of research.[29] The 2012 policy report *Responsible Conduct in the Global Research Enterprise*, produced as a cooperative project with the InterAcademy Council (IAC), offers an international consensus statement on the meaning of responsible conduct and the way to promote it (IAC and IAP, 2012). The report concludes that:

- Researchers have the primary responsibility for maintaining standards of responsible research and should agree on the standards to be observed in multidisciplinary collaborations.
- Research institutions should develop clear definitions and rules about responsible conduct and foster an environment of integrity, including the establishment of effective mechanisms for addressing allegations of misconduct.
- Institutions and agencies should support responsible, high-quality work through funding practices that emphasize quality over quantity of results.
- Journals and investigators should publish only original material.

[29] IAP was founded in 1993 to help national science academies advise their respective national policymakers on global scientific issues; for more information, see www.interacademies.net/. The InterAcademy Council (IAC) is an IAP Observer organization, established in 2000 as a source of expert scientific advice for global organizations such as the United Nations. For further information about IAC, see *www.interacademycouncil.net/*.

INCORPORATING STANDARDS OF BIOSAFETY, BIOSECURITY, AND DUAL USE RESEARCH INTO THE INTERNATIONAL EDUCATION INITIATIVES ON RESEARCH INTEGRITY AND RESPONSIBLE CONDUCT OF RESEARCH

Emerging ethical questions and standards of practice in biosafety, biosecurity, and research with dual use potential fit readily into the broad spectrum of issues addressed in RCR education. Three of NIH's recently defined areas for RCR education are directly relevant to biosafety and dual use issues: policies on safe laboratory practices, the scientist as a responsible member of society, and the social and environmental impacts of research.[30] Moreover, as the cases discussed in Chapter 4 illustrate, relevant issues arise in the majority of the broader core areas. For example, RCR education in the life sciences can readily address the following topics:

- Mentors' responsibility for ensuring that trainees work safely in the laboratory, and trainees' responsibility for learning and practicing safe laboratory and clinical methods, asking for guidance when they feel unsure, and reporting spills and exposures;
- The secure collection, documentation, and management of research data, and policies, regulations, and best practices regarding ownership and sharing of data and research tools with dual use potential;
- How collaborative research, particularly across national borders, is governed by national and international regulatory standards on the shipping of materials, and how export controls define security interests and threats;
- How the standards of open publication of study design and methods as well as research results and interpretation may present challenges for work that explores novel infections or techniques with dual use potential; and
- The growing role of scientific and ethical peer review in decisions about funding for research with dual use potential and publication of its results.

An additional approach to teaching about biosafety, biosecurity, and dual use issues as part of RCR education has been advocated by proponents of a researcher's code of ethics. In 2005, the U.S. National Science Advisory Board for Biosecurity (NSABB) recommended developing a "Code of Ethics for Life Scientists," and in 2007 published guidelines for developing a code of conduct for dual use research (NSABB, 2007). NSABB later outlined core professional responsibilities and general research responsibilities that could be incorporated into a code of conduct related to dual use research in the life sciences (NSABB, 2010). Several prominent life scientists and science policy scholars have also proposed a "Hippocratic Oath for scientists" that both students and established investigators can use as a point of reference for professional behavior (Rotblat, 1999; Jones, 2007; Cressey, 2007; Lehn, 2011). Most such codes incorporate a provision against doing harm through research, which would prompt reflection on research integrity as well as dual use potential.

As noted in Chapter 1, a strong theme for education and outreach related to dual use issues is to treat the topic within a broader framework of responsible conduct of research. Examples of how this framing works in practice may be

[30] These topics have been addressed, for example, during discussions at meetings of the Biological Weapons Convention.

found in the activities of the IAP Biosecurity Working Group, established in 2004 to undertake IAP's work at the intersection of science and security, with a focus on dual use issues.[31] From the beginning, the group couched its work in the context of responsible conduct of science and the social responsibility of science. The group's first product, the 2005 IAP Statement on Biosecurity, identified "fundamental issues that should be taken into account when formulating codes of conduct" (IAP, 2005). In cooperation with other international scientific organizations, the group organized the 1st and 2nd International Forums on Biosecurity in 2005 and 2008, respectively; education and codes of conduct were discussed in both meetings.[32] These activities led to the State Department's request to hold the workshop on *Challenges and Opportunities for Education About Dual Use Issues in the Life Sciences*, which in turn led to the project that is the subject of this report.

---

[31] The current membership includes the national academies of Australia, China, Cuba, Egypt, India, Nigeria, Poland (chair), Pakistan, Russia, the United States, and the United Kingdom.

[32] The first forum did not produce a report, although the agenda and participants list are available at *http://nas-sites.org/biosecurity/international/*; the report of the second forum was produced by the National Research Council (2009d).

Chapter 3
The Science of Learning

This chapter offers a brief primer on the concept of active learning, summarizing the growing research base and introducing its applications in a variety of educational settings.[33] After outlining some of the general characteristics of common approaches to faculty development programs, it then describes a program developed by the National Academies to apply the concepts in an effort to improve undergraduate biology education. As mentioned in Chapter 1, that National Academies program is the model for a new international project to develop networks of life sciences faculty able to apply active learning methods to responsible conduct and dual use issues. Chapter 4 will repeat much of the basic material presented in this chapter, but in the context of how it was presented and modeled in a real learning situation.

Methods for active learning instruction have been under development and refinement for more than 130 years. A large and growing body of evidence, cutting across scientific disciplines, is demonstrating that modern versions of these methods offer the potential for significantly improved learning in comparison to traditional, student-passive, lecture-based instruction (NRC, 2000; Handelsman et al. 2007; Knight and Wood, 2005; Prince, 2004; NRC, 2011d; Meltzer and Thornton, 2012). A common feature of active learning instruction is that it involves students in their own learning more deeply and more intensely than does traditional instruction. In all cases, the instructional methods (1) are based on, assessed by, and validated through research on teaching and learning, (2) incorporate classroom and/or other activities that require all students to express their thinking through speaking, writing, or other actions that go beyond listening and taking notes, and (3) have been tested repeatedly in actual classroom settings and have resulted in objective evidence of improved learning. Learner-centered environments are more likely to be collaborative, inquiry based, and relevant (Brewer and Smith, 2011). The research suggests that there are many teaching strategies that can support active learning. These range from problem-solving/discussion sessions in class to original investigations that may be student designed. Table 3-1 contains descriptions of a variety of active learning techniques, with illustrations of how they might be used in biology classes.

The methodology has been effective in various settings, from small groups to large lecture-based courses.[34] At the college level,

[33] Many of the terms associated with active learning are defined in the Glossary.

[34] Engaging students in active learning in large class settings such as lecture halls has garnered much attention from education researchers. A number of techniques, including the use of individual wireless response systems (clickers) that allow students to answer questions anonymously, "think-pair-share" techniques in which students develop their own answers to questions and then discuss their answers with a student next to them (often combined with clicker questions), and similar exercises involving peer learning and engagement have proven to be valuable active learning tools in these kinds of settings. For additional

formal active learning has been used in courses that range from introductory undergraduate to graduate level. The data show that there is no significant difference in the positive results achieved by predominately female or male, and heterogeneous or mixed gender groups. However, the positive effect of small-group learning was significantly greater for groups composed primarily of African American and Latino students compared with predominantly Caucasian and relatively heterogeneous groups (Springer et al., 1997). Additionally, workshops for teachers and college and university faculty increasingly use active learning methods.

It has been demonstrated that to be well understood, factual knowledge must be put in a suitable conceptual framework. The data show that framing learning in the sciences as four intertwined strands of proficiency provides a sound basis for creating effective teaching and learning experiences at all levels (NRC, 2007b, 2011d); these are:

- understanding scientific explanations,
- generating scientific evidence,
- reflecting on scientific knowledge, and
- participating productively in science.

A critically important aspect of effective instruction is the integration of learning about process and content. Although this is not always the case in practice, science teaching laboratories historically have been viewed by many faculty as the place to provide valuable and unique opportunities for the learner to engage in conceptual materials. Rather than being viewed as an add-on or distraction from content mastery, the laboratory is one of the many pathways to factual knowledge and deeper conceptual understanding (NRC, 2006b). However, the science education community is now beginning to view the entire course (classroom, laboratory, and field experiences), especially at the introductory level, as an opportunity to integrate content with scientific processes and skills and to help students understand and appreciate the relevance of science to their own lives and that of their communities (Labov, 2004; Handelsman et al., 2006; AAAS, 2011; PCAST, 2012; NRC, 2012a). Critical reflection, as called for in the third strand, is an essential component of virtually all effective approaches to learning. To date, this is the only practice that has demonstrated student learning gains in understanding the nature of science (NRC, 2006b, 2008). Reflection provides students with the opportunity to explore their level of understanding with other learners (and the teacher) and helps them become more aware of their own levels of learning. Students become able to self-monitor their learning, they plan and set goals, and they have many opportunities to reflect on their learning and adapt as necessary. The value of such "metacognition," or self-monitoring of one's learning, has been demonstrated by many studies and is a critical component of effective teaching and learning strategies (Zimmerman and Schunk, 2001). Active learning, properly implemented, encourages metacognition. Given the complexities of the ethical and social dimensions in the responsible conduct of science, it is also important to include time for various forms of reflection throughout a course.

Research shows that understanding is built on a foundation of existing conceptual frameworks and experiences. While prior knowledge can support further learning, it may also lead to pre- or misconceptions that act as barriers to learning. Prior understandings are influenced by culture, which has implications for

---

information about encouraging active learning in large class settings for different disciplines, see for example MacGregor et al., 2000; Allen and Tanner, 2005; Caldwell, 2007; Poirier and Feldman, 2007; Stranger-Hall et al., 2010; Wood and Tanner, 2012)

TABLE 3-1 Examples of Learning Objectives and Active Learning Techniques

Biology Example and Instructions	Objectives
Brainstorming	
Answering the following question in large group. One person records answers. Optional: Arrange the list into two or more categories (*e.g.*, abiotic vs. biotic factors) *Question:* What does a plant need to survive	Brainstorming elicits responses from large audience and aggregates them into a single list. It provides the instructor and students with an overview of the group's collective knowledge. By separating the brainstorm list into two or more categories, students evaluate how well they understand the role of each response in a specific context.
Case study and decision making	
Read the following case. Write a paragraph to explain what the patient should do next. Justify your recommendation with biological reasons. *Case:* A patient expressed eye irritation, which the doctor diagnosed as conjunctivitis. Antibiotic treatment alleviated the symptoms within a few days, but the symptoms returned two weeks later. The doctor recommended taking antibiotics again.	Cases engage students in solving a problem in a real-life context. To solve them, students need to evaluate what they know about infectious disease, causal agents, and antibiotic resistance; apply that knowledge to the case; and determine what additional information is needed to make a recommendation.
"Clicker" questions	
Answer the following question on your electronic response keypad. *Question:* Which organisms are most distantly related? (a) bacteria and archaea; (b) plants and animals; (c) plants and fungi; (d) humans and fungi	Clicker questions require students to gauge whether they understand a concept or topic, thereby engaging students in the ensuing activities (*e.g.*, lecture) about that topic.
Group exams	
Work with a group to discuss the following statement. Write your answer individually. *Statement:* Explain the role of aflatoxin in liver cancer.	Group exams engage students in working collaboratively to identify creative solutions to a problem. Writing individual answers requires students to evaluate how well they understand the topic and its underlying concepts.

Mini-map	
Arrange the following terms in logical order. Explain (using arrows or words) how the terms relate to each other. *Terms*: tRNA, DNA, protein, mRNA, amino acid, translation, transcription, replication, promoter	Mini-maps engage students in developing a non-verbal representation of a concept. The process of developing a visual arrangement requires students to evaluate different ways that terms can relate to each other and to appreciate that a biological process may not be unidirectional or linear.
One-minute paper	
Write for one minute to answer the following question. *Question:* What about the structure of DNA suggests a mechanism for replication?	One-minute papers engage students in articulating their knowledge about a topic or applying their knowledge to another situation. By writing their answer in one minute, students need to evaluate the most important and relevant components of their argument.
Pre/post questions	
Write for one minute at the beginning and end of class in response to the following statement. Explain any differences between your responses. *Statement:* Describe two mechanisms that a bacterium can use to harm a plant.	Pre/post questions can take many forms, including one-minute papers or clicker questions. They engage students in thinking critically about a specific question or problem. By comparing pre/post responses, students evaluate whether and why their answers changed during the class period.
Strip sequence	
Use your textbook as a guide and work with a partner. You write the important steps in meiosis; your partner writes the important steps in mitosis. Cut the steps apart and scramble the order. Each of you should try to put the other person's steps into the correct order. Discuss.	Strip sequences engage students in recognizing cause and effect and in determining the logical sequence of events. When students derive their own strip sequences, they need to evaluate the critical steps in the process.
Statement correction	
Discuss with a partner what is wrong with the following statement. Propose an alternative statement that is correct. *Statement:* "I don't want to eat any viruses or bacteria, so I refuse to buy foods that have been genetically modified."	Statement corrections engage students in evaluating what concepts are misrepresented and in determining what information they need to correct it.

SOURCE: From *Scientific Teaching* by Handelsman, Miller, and Pfund. Copyright © 2007 by W.H. Freeman and Company. Used with permission.

the development of curricular materials that may be used to teach responsible conduct of research for international audiences (NRC, 2008). The importance of engaging learners' prior understanding as they learn new material is an important insight from the science of learning (summarized in NRC, 2000).

Faculties are adept at designing curricula to engage students in key scientific practices: talk and argument, modeling and representation, and learning from investigations (NRC, 2008). They are less facile at course design with active learning as a goal. Most instructors first select the textbook, then compile the course syllabus and assignments, construct the examinations, and finally describe learning goals and objectives. Active learning courses are best designed when the first step is the identification of goals and objectives and then the syllabus. This "backward design" process (Wiggins and McTighe, 2005), also called reverse design, is intended to ensure that learning objectives inform instructional and assessment strategies through explicit articulation of these two critical components of the learning process and then integrate them into the design of the course at the outset.

Assessment of student learning should be both formative and summative. Formative assessment is generally low stakes (either none or a small portion of the student's grade) and is used regularly throughout the learning process, providing feedback to both students and faculty about student learning and academic progress. Summative assessment, conducted at the end of the block or course, provides information about student learning gains and the overall success of the effort. Both formative and summative assessments should be used for subsequent course/curriculum restructuring. Without assessment that is closely aligned to learning objectives, it is difficult to determine the effectiveness of the curriculum.

Even though much of the research cited and the examples referenced above have occurred in the United States, a growing number of countries are undertaking efforts to reform and transform the way that science is taught. Collaborations between U.S. and non-U.S. universities are assessing the effectiveness of active learning in a variety of contexts. A study conducted simultaneously in Sweden and the United States suggests that curricula that actively engage the student do appear to make a permanent change in their conceptual framework. As long as 2½ years after the instruction, students had a "good" grasp of concepts (Bernhard, 2001). A review of the literature finds there is broad but uneven support for the core elements of active learning (Prince, 2004). "Students who learn in small groups generally demonstrate greater academic achievement, express more favorable attitudes toward learning," and remain enrolled in science, technology, engineering, and mathematics (STEM) courses and programs "to a greater extent than their more traditionally taught counterparts" (Springer et al.,1997:42).

Conferences of international scientific unions and other professional organizations now routinely include sessions that feature symposia, workshops, or other sessions that emphasize teaching and learning. The International Brain Research Organization (IBRO), a global network for neuroscience research, organizes "Teaching Tools Workshops" that assist African countries in adding or improving the teaching of neuroscience. The workshops include both content and teaching methods, with a strong focus on learner-centered approaches.[35] The 2012 Lilly Conference on College and University Teaching, which draws participants from the

---

[35] Further information is available at http://dels-old.nas.edu/USNC-IBRO-USCRC/activities_workshops.shtml#past.

United States and overseas, chose the theme Evidence-Based Learning and Teaching to reflect that approaches to teaching and learning should be based on scholarly activity.[36] Additionally, the IEEE International Conference on Teaching, Assessment, and Learning for Engineering (TALE) is held each year in the Asia-Pacific region and complements the Frontiers in North America and the EDUCON in Europe/Middle East/Africa conferences.[37] At the primary and secondary level, IAP–The Global Network of Science Academies, has promoted what it calls "Inquiry-Based Science Education" since 2001 through activities led by the Chilean Academy of Sciences.[38] The next section describes the model used for the project that is the subject of this report.

PUTTING RESEARCH INTO PRACTICE[39]

Introductory science courses at large universities in the United States serve as the portals that connect undergraduates to frontiers in research and scientific ways of thinking. An introductory undergraduate biology course might be the only exposure many students have to the life sciences, or to any of the sciences. It often serves as the best opportunity to interest students in a biomedical research or other life science careers.

According to the 2003 National Academies report *Bio2010: Transforming Undergraduate Education for Future Research Biologists*, however, teaching practices have not kept pace with advances in scientific research about learning (NRC, 2003). Consequently, the gateway through which most students pass is antiquated, misrepresents the interdisciplinary, collaborative, evidence-based culture of science, and fails to implement current knowledge about how people learn. *Bio2010* identified faculty development as a crucial component in improving undergraduate biology education and the authoring committee suggested the development of a "Summer Institute" to bring life sciences faculty together to work on improving education. This Summer Institute would focus on integrating current scientific research and appropriate pedagogical approaches to create courses that actively engage students in the ways that scientists think. The committee further recognized the need for ongoing reinforcement of teacher development and the benefits of interactive activities to produce participants who would be fully able to use their new pedagogy and content knowledge effectively.

Characteristics of Faculty Development Programs

Over the years, dozens of programs across all the STEM disciplines have been implemented to build the capacity of faculty to teach effectively. They are a subset of the more general category of "train-the-trainer" programs in which more experienced educators seek to impart knowledge or skills in a way that can be sustained after the initial encounter. The newest programs, such as some of those described in this report, draw on

---

[36] For further information, see *http://cml.esc.edu /node/629.*

[37] The Institute of Electrical and Electronics Engineers, a professional association headquartered in New York City with more than 400,000 members in more than 160 countries, now uses IEEE for everything but formal, legal matters. For further information, see *www.tale-conference.org/ tale2013/venue.php*.

[38] For further information, see www. interacademies.net/Activities/Projects/12250.aspx.

[39] A version of the text in this section appeared in the letter report of the planning meeting for this project, Research in the Life Sciences with Dual Use Potential: An International Faculty Development Project on Education about the Responsible Conduct of Science (NRC 2011e:14-19). The material has been lightly edited and updated to reflect developments since the meeting. The section entitled "Characteristics of Faculty Development Programs" is new material prepared for this report.

the science of learning to inform the faculty development programs themselves, infusing the workshops/meetings/ institutes with active learning principles and practices. A report released in 2013 on *The Role of Scientific Societies in STEM Faculty Workshops* (Hilborn, 2013), for example, provides descriptions and initial assessments of a number of programs run by major U.S. professional societies. Although the programs vary in terms of a number of features, such as their target audiences (e.g., junior versus senior faculty, type of institution), length (e.g., from a weekend to a week to 10 days), and location (e.g., standalone or as part of a professional society meeting), they also share a number of major characteristics.

- Simply stated, the goals of all the STEM faculty programs discussed here are to develop expert competence in teaching, to enhance faculty views of teaching as a scholarly activity, and to promote the use of evidence in evaluating the effectiveness of teaching practices.
- All of the initiatives promote, either explicitly or implicitly, the importance of "scientific teaching."
- The meetings generally consist of a mix of plenary sessions, often carried out with interactive engagement techniques—to model what the leaders hope the participants will implement in their home institutions—and smaller breakout and discussion sessions.
- While many effective pedagogical practices cut across disciplines, their effective implementation requires broad knowledge of the discipline and its modes of discussion and argument. Hence, all of the programs described here have the participants think about (and in some cases practice) effective pedagogical methods within the context of the discipline. This method builds on the content knowledge of the participants and prepares them more directly for the teaching decisions they will need to make in their own classrooms.
- ...all of the program leaders recognize that a one-time workshop is unlikely to produce the kind of expert teaching competence required of an effective instructor. The programs use a variety of mechanisms to continue interactions among the participants (peer mentoring and coaching) and with the program leaders. (Hilborn, 2013:6-9)

Together, these and other programs offer a number of different models for undertaking faculty development.

One Model in Detail: The National Academies Summer Institutes in Undergraduate Education in Biology (NASI)

One substantive result of the recommendation in *BIO2010* was the development of the annual National Academies Summer Institute for Undergraduate Biology Education (NASI).[40] This institute is designed to model the scientific teaching principles on which it is founded and draws on the expertise of both participants and presenters.

NASI provides a venue each year for teams of faculty from primarily research-intensive universities to meet for five days of in-depth discussions, demonstrations, and working sessions on research-based approaches to undergraduate biology education. The idea is to generate the same atmosphere as a Gordon Conference or a Cold Spring Harbor research course, but with the topics being issues in education rather than, for instance, bacteriophage genetics. Current research in effective pedagogical practices in undergraduate science education, active learning, assessment,

[40] For additional information see *http://academiessummerinstitute.org* and an article by Pfund et al. (2009).

and diversity are woven throughout the week, creating a forum for participants to share ideas and develop innovative instructional materials that they are expected to implement when they return to their own campuses.

Initiated with a pilot institute in 2003, NASI convened annually during the last week of June on the campus of the University of Wisconsin, Madison from 2004 to 2011. The target audiences have been faculty and academic leaders from universities where large courses, especially at the beginners' level for both life sciences majors and for students with other career goals, provide significant impediments to reform. Most universities have sent a team of 2-3 people to one institute. Others have sent multiple teams (of different people each year) over two or more years. NASI has been supported primarily through funding from the Howard Hughes Medical Institute (HHMI; through summer 2011) with additional support from Research Corporation for Scientific Advancement and the Burroughs Wellcome Fund.

Based on NASI's success, HHMI provided a new award to the program that has enabled its expansion to up to eight institutes each year in regions across the United States. Four regional institutes were organized in 2011, seven in 2012, and another seven in 2013.[41] These regional institutes adhere to the structure and emphasis of the Madison NASI but also expand the pool of educators beyond faculty in research-intensive universities, participants (e.g., graduate students and postdoctoral fellows in addition to junior and senior faculty), and areas of expertise (beyond a primary focus on the biological sciences). Data about the participants in these institutes and how they change their approaches to teaching and student learning will continue to be collected and analyzed.

Participants at the Madison NASI were selected based on a rigorous application process overseen by a National Academies committee; applications for the regional institutes are monitored by the local organizing committees using similar criteria. There is a particular emphasis on including pretenured as well as senior faculty as members of the team. NASI also trains a cadre of mentor/facilitators who work with participating teams each summer. Many of these facilitators are NASI alumni, selected for this honor based on observations of their performance during the institute they attended.

Although an individual regional institute may reorganize the schedule to some extent, each institute typically consists of a series of plenary sessions in the mornings and facilitated small group activities during the afternoons. All plenary sessions model the kinds of evidence-based active teaching and learning that the Institute stresses for improving undergraduate education. Topics include subjects such as active teaching, how people learn, formative and summative assessment, teaching to diverse student populations, mentoring, and working with colleagues to improve teaching and learning.

Each small group typically consists of participants from three university teams and focuses on producing a "teachable tidbit" in some broad area of biology or interconnected disciplines (e.g. biology/chemistry, biology/mathematics). A tidbit is a module that integrates aspects of classroom, laboratory, or field experiences, assessment, and techniques to help diverse student populations learn more effectively. Small groups are given time to interact with each other during the week to critique each other's tidbits as they are developed. Each team then presents its "tidbit"

[41] Links to general information about the regional institutes and the dates and locations of regional institutes in a given year are available at *http://academiessummerinstitute.org*.

on the next-to-last day. Each tidbit is peer-reviewed by other participants, facilitators, and members of the organizing committee.

All resources and products of each NASI are collected on an Academies portal and made available to all participants, current and previous.

Over the course of the NASI program (2004-2012) 710 people have participated from 167 institutions in 46 states and the District of Columbia. Because so many of these participants serve as instructors in large lecture-style courses, collectively they have taught more than 250,000 undergraduates.

The National Academies recognizes the commitment of these participants by naming each an "Education Fellow in the Life Sciences" for the year following their attendance at NASI. Participants also identify key academic leaders on their campuses who are notified about the honor.

From its inception, NASI has also been a research project. Self-reported data from participants are collected and analyzed regularly to determine the impact of this initiative (e.g., Pfund et al., 2009). In addition, HHMI sponsors a midyear meeting for one representative from each university team approximately 6 months after their NASI participation to measure success, challenges, and new activities that have emerged from their participation. The data and information gained are used in a constant process of adjustments and iterations to improve the NASI; the current version bears only a modest resemblance to the original institutes. This commitment to continuous assessment and adjustment as needed for faculty and students as well as courses and programs is another hallmark of active learning. Chapter 4 describes how the lessons of active learning and the NASI approach and experience were applied to new material in a new setting.

Chapter 4
The Institute

This chapter describes the preparations for and activities during the Educational Institute for Responsible Research on Infectious Diseases: Ensuring Safe Science in the 21st Century (hereafter, the Institute), which was held in Aqaba, Jordan, in September 2012. As discussed briefly in earlier chapters, the Institute applies a model developed by the U.S. National Academies (the National Academies Summer Institutes in Undergraduate Biology Education, or NASI) to use active learning methods to improve the quality of undergraduate biology education to the challenges of creating networks of faculty able to teach about dual use issues in the context of responsible conduct of science. The choice of NASI as the model from among the many available approaches to faculty development programs (see Chapter 3) reflects the knowledge and experience that the Academies have accumulated in a decade of conducting them, the data that the project has collected and continues to collect about its efficacy (e.g., Pfund et al., 2009), and the fact that some members of the Institute's organizing committee were selected because of their leadership in NASI to get the project off the ground. The project is a collaboration among the National Academies, the Bibliotheca Alexandrina, and The World Academy of Sciences (TWAS). The material in this chapter on active learning methods provides an opportunity to show how the concepts introduced in Chapter 3 can be presented and applied in an actual learning situation.

THE PLANNING MEETING AND PILOT

The original development of NASI included a pilot test of the design. Plans for a similar, smaller-scale pilot were included in the grant for the Middle East–North Africa (MENA) project. The insights from the pilot, carried out as part of the planning meeting held at TWAS in Trieste, Italy, in early June 2011, were essential to the development of the Institute. Experts from Europe, the United States, Egypt, and South Africa joined the members of the National Academies committee overseeing the project.[42]

The Trieste meeting built on the Warsaw workshop's strong emphasis on active learning approaches to teaching and the inclusion of experts in pedagogy along with experts in dual use issues, responsible conduct of research, and various fields of relevant life sciences research (NRC, 2011c). The initial discussion of the project's goals and fundamental concepts focused on dual use and responsible conduct of research and was followed by examples of general life sciences education, as well as "train-the-trainer" programs that make use of active learning methods. In addition to discussions, attendees had a chance to engage actively with some of the methods themselves. In particular, small groups of attendees were given the task of setting general goals and specific learning objectives for the Institute. The results of the

[42] The list of planning meeting participants is shown in a 2011 letter report (NRC 2011e).

small groups' deliberations provided the foundation for the final day's general discussion of next steps, which served as the basis for the committee's conclusions about the overall design of the Institute, which were presented in a letter report (NRC, 2011e). The conclusions were intended

> to serve as global guidelines applicable…to any country wishing to adopt this educational model that combines principles of active learning and training with attention to norms of responsible science. It aims to address the unmet need of respectfully incorporating into existing science teaching and research (especially in the field of emerging infectious diseases) the ideas of conducting science responsibly, of cultivating a culture of laboratory safety, and of raising awareness within the local scientific community of the consequences of misusing research with dual use potential (NSABB, 2008; NRC 2009c). (NRC, 2011e:10)

Five general considerations were identified to frame the Institute:

- Responsible conduct of research/research integrity as core themes.
- The importance of respecting and adapting to the national context of the workshops' host countries.
- The advantages of the science of learning and scientific teaching approach.
- The value of creating networks of faculty and institutional support for the sustainability of efforts.
- Essential role of assessment and evaluation. (NRC, 2011e:12-14)

The full text of the conclusions is worth quoting at length (see Box 4-1) because of their influence on the development and implementation of the project's next phases. The letter report also includes discussions of the detailed lessons that the meeting provided for the design of the Institute. That text is provided in Appendix C. The actual work of designing the Institute is described in the next section.

DESIGNING THE INSTITUTE

The design of the Institute followed the steps outlined in the planning meeting in Trieste and described in the letter report. The committee members formed three subgroups to (1) design the content, (2) develop the pedagogy, and (3) review and evaluate the applications from prospective participants. The content and pedagogical elements were chosen to support the implementation of the Institute's goals as formulated in Trieste to cultivate future leaders in responsible science and research integrity (NRC, 2011e:17). For participants unfamiliar with the ethical and legal responsibilities of physical and life scientists or issues in the responsible conduct of science, the Institute would provide an introduction. For those who had experience with these topics, the Institute would provide an opportunity to gain a deeper appreciation and share their insights. Since science faculty in many parts of the world receive little formal training in teaching or knowledge of the emerging scientific research on human learning and cognition that can help to improve pedagogy (e.g., NRC, 2000), the committee anticipated that participants' familiarity and experience with active learning techniques would be equally varied. The committee believed it was important to provide some of the basics of best teaching practices, as supported by cognitive science and discipline-based education research, in addition to the scientific and ethical aspects of responsible science. To help Institute participants better understand elements of responsible conduct of research (RCR), responsible science (RS), and best practices in pedagogy, the committee created pairs of content and active learning techniques. Definitions of these techniques are provided in the Glossary; further descriptions of the active learning techniques with examples from biology of how they can be applied are shown in Table 3-1, and additional resources are in Appendix D.

The Schedule at a Glance (Figure 4-1) shows the thematic and chronological architecture of the Institute, which emulates the design of NASI. A facilitator-training day preceded the initiation of the Institute. Over the course of 5½ days, the participants took part in morning content/pedagogy sessions and spent the afternoons and evenings in small groups to develop teaching modules based on selected topics and using active learning tools. These modules were presented to, and discussed by, the faculty and participants. On the last day, the participants met together by country and presented their ideas for implementing the Institute's content and pedagogy in their home institutions and countries to everyone attending the Institute.

Recruitment of Participants

A crucial action identified at the planning meeting at TWAS was the need to engage early in strategic discussions about supporting the cohort of participants in the Institute upon their return to their home institutions (NRC, 2011e:15). Using the extensive communication networks of the Bibliotheca Alexandrina as well as its experience in managing competitive application processes, teams of participants from different MENA countries were invited to apply. In addition to their academic accomplishments, applicants were judged on the basis of a personal statement, which elaborated their individual teaching philosophy, the types of courses they taught, what they each hoped to achieve by attending the Institute, and their contributions to science. The application is shown in Appendix G. A total of 56 applications from qualified individuals were received. Of those, the staff in consultation with the committee selected 32 participants, 28 of whom attended.

The Institute in Aqaba was attended by a cross-section of individuals from the MENA region: Algeria (4), Egypt (14), Jordan (3), Libya (1), and Yemen (6), of whom 8 were women and 11 were part of a team or from the same institution. Based on the experience of NASI, teams were encouraged to include at least one senior faculty member. The list of participants and their affiliations is in Appendix F.

Facilitator Preparation

An integral aspect of NASI is the preparation of its facilitators and frequent opportunities for them to work together each day to address problems and develop solutions to those problems collectively. Initial preparation of facilitators takes place the day preceding each NASI. The preparation sessions help new facilitators recognize the difference between teaching and facilitation and allow them to practice strategies to maintain their roles as facilitators rather than educators (Table 4-1). As with all aspects of NASI, facilitator preparation draws on findings from the sociological and organizational research literature on group dynamics. Introducing facilitators to this body of work, which focuses on various stages of group formation and cohesion (e.g., Tuckman, 1965; Richards and Moger, 1999; Stetson, 2009) and encourages them to share this information with the members of their groups, can help individual group members recognize their strengths and weaknesses associated with group work. Recognizing these factors can, in turn, help participants to better facilitate group situations

BOX 4-1
Conclusions from the Planning Meeting[a]

Responsible conduct of research/research integrity as core themes. Building on a prominent theme from the Warsaw workshop and other NRC reports about education related to dual use issues (NRC, 2004, 2009b, 2011c), broader principles of responsible conduct and research integrity rather than the "dual use" theme were chosen as the foundation for faculty development. By embedding the EPI [Egyptian Prototype Institute] in general discussions on professional conduct, participants accepted the idea that this more general approach would likely be more enduring and sustainable than focusing only on dual use issues. It also resonated with the participants from Egypt for whom a more comprehensive framework beyond research with dangerous pathogens is a more realistic educational opportunity. Such an inclusive approach would also enable future workshops to take advantage of other initiatives....

Importance of respecting and adapting to the national context of workshop host countries. One of the insights from earlier efforts to develop education programs on responsible conduct of science and dual use issues is the wide variation in higher education structure and process, and national education policy and how those differences could affect the design and implementation of programs (NRC, 2011c; Rappert, 2010).

- The difficulty of introducing new material, especially beyond core science topics, into crowded curricula is a common concern among nations. In some countries introducing entire new courses into existing curricula can have a direct impact on the development and implementation of faculty networks both at an institutional and national level and efforts to develop nationwide approaches may be difficult. In some countries where institutions of higher education are largely autonomous (e.g., the United States), development of new courses can essentially result from an instructor's initiative, with only limited approval needed from immediate supervisors. In nations with a centralized ministry of higher education (e.g., Egypt) a new course could require approval by national authorities, an often lengthy process.
- One of the most sensitive areas for teaching about dual use and related issues is the political and historical context of different countries, which in some cases may make faculty reluctant to become involved in anything associated with "security." This supports the point already made above about the advantages of embedding dual use issues within the broader framework of responsible conduct. It also may affect the choice of the local partners, for example, understanding whether formal or informal endorsement by certain government or education officials is essential or how important it might be to work with an institution that by virtue of its prestige or connections can provide flexibility for teaching new courses for its faculty.

- The importance of local context for the successful design of a faculty development program underscores the need of a preparatory site visit(s) as part of the planning process. One outcome of the Trieste workshop was the decision to send a small team of staff and Committee members to Egypt to meet with local faculty, university officials, and government administrators in Fall 2011. The purpose of these meetings is to inform university and government leaders about the planned workshop, and acquire their active support for its successful execution, for the participation of junior faculty, for any follow-on activities originating from the participants, and for the initiation of a network of faculty-workshop participants who will subsequently become trainers for other faculty and their students. An important point to discuss will be the mechanism by which the participants will be chosen so that local mechanisms will be considered. As mentioned in the previous bullet, the advice of well-chosen local partners is invaluable in understanding the political sensitivities and planning a successful visit.

Advantages of a "science of learning" approach. The enthusiasm among participants for their experience with active learning reinforced the message from the Warsaw workshop about the value of such approaches in education about dual use and related, broader issues. The relevance of adopting such methods for classrooms and laboratories across the world is supported by the decision by the World Health Organization to revamp its biosafety train-the-trainer programs to adopt similar active learning methods (WHO, 2006, 2010).

Sustainability of efforts: Value of a network approach and institutional support. As already mentioned, a continuing challenge for efforts to promote new concepts, materials, and pedagogical approaches is the competition for space in a crowded curriculum. It is essential that, from the beginning, the planning for any such effort include a focus on strategies to make the project sustainable. The lessons from efforts in many other areas reinforce the value of building networks of faculty who can share experiences and provide mutual reinforcement (NRC 2010). For example, creating opportunities for participants in a faculty development workshop to get together after their initial experience in implementing what they have learned has proved extremely valuable to sustaining commitment and momentum (Pfund et al., 2009). In a broader context, building institutional support for sustaining not only the network but the faculty's ability to introduce others to these concepts as well as support for both teaching and research would help foster the culture of responsible science.

Assessment and evaluation. The "science of learning" approach emphasizes concrete goals and continual, measurable outcomes of student performance, whether qualitative or quantitative. Effective evaluation depends on incorporating assessment as an integral part of the follow-on activities and as such would inform any strategies to sustain these educational efforts.

[a] This text is reproduced from NRC, 2011e:12-14.

TABLE 4-1 Characteristics of effective teaching compared with effective facilitating

Effective Teaching	Effective Facilitation
Emphasizes learning by individuals but also can foster collective learning.	Emphasizes collective learning.
Imparts knowledge and conceptual frameworks to students.	Guides the processes for the development of knowledge and skills.
Emphasizes new knowledge acquisition and understanding in specific content domains but also helps students understand the need for reflecting on their learning.	Emphasizes processes of learning, reflection about learning, and new, deeper understanding of preexisting knowledge from many domains (e.g., personal, professional).
Teacher often serves as the knowledge expert.	Knowledge and expertise are shared among the facilitator and other learners.

Source: The table has been adapted and modified from Miller and Pfund (2011).

among their own colleagues or students, as elaborated in Table 4-1.

Given the success of these strategies and activities and the fact that most facilitators at the Institute had no experience with this role, two members of the planning committee and one staff member for this project, who have been associated with NASI many years, organized a preparatory session for all the facilitators of the Institute. Facilitation goes beyond the use of good teaching practices and, as noted in a recent edition of the facilitation manual that was developed for NASI, "Effective facilitation is a nuanced balance of leadership and participation, assembly and deconstruction—each of which can (and should) be practiced" (Miller and Pfund, 2011:3). Additional differences between teaching and facilitating are described in Table 4-1.

In these sessions, facilitators also learned about understanding and dealing with different interpersonal relationships and conflicts that often develop among group members through the course of an Institute. These kinds of dynamics include:

- *Respecting each member of the group and her/his contributions.* That is, effective listening as well as talking, respectful questioning of statements or opinions offered by group members, nonjudgmental discussions and interactions.
- *Keeping the group focused on the task at hand.* Understanding the difference between relevant tangents and those that lead the group away from their goals and tasks. *Using time thoughtfully.* Although facilitators agreed each afternoon on the goals and work to be accomplished the following day, effective facilitators recognize differences in group dynamics (these differences include those between groups as well as those within groups that might develop over time as group members have additional opportunities to interact with each other). In some cases, slowing down the pace of work is important while in others the group will be able to work more quickly than anticipated. Rigid adherence to a schedule that is designed prior to an Institute could interfere with actual progress

FRIDAY SEPT 7 FACILITATOR TRAINING DAY		SATURDAY SEPT 8		SUNDAY SEPT 9	MONDAY SEPT 10	TUESDAY SEPT 11	WEDNESDAY SEPT 12	THURSDAY SEPT 13
		BREAKFAST 07:00						
BREAKFAST 08:00	08:00	Theme 1: The development of professionalism in science		Theme 2: Conducting research responsibly	Theme 3: Being part of the responsible scientific community		GROUPS PREPARE FOR PRESENTATIONS	NEXT STEPS
Training 09:00		PARTICIPANTS Careers in science: Pathways and inspirations		CONTENT 2 - Research with animals and human subjects (2a) Introducing a new virus in the field (2b) The Guatemala syphilis studies	CONTENT 3 - Collaborative research; Authorship and publications; Peer review (3a) The Darsee affair (3b) Who is an author?	CONTENT 4 - Research with dual use potential (4) Powerpoint discussion		
		COFFEE BREAK 10:15						DISCUSSION AND ADJOURNMENT 10:00
	10:30	HIGHER EDUCATION in the MENA region: Country focus		PEDAGOGY 2 - Assessment and introduction to Active Learning	PEDAGOGY 3 - Active learning	CONTENT 5 - Biosafety concerns in research (5) Studies in H5N1 influenza virulence	GROUP PRESENTATIONS	COMMITTEE AND FACILITATOR DEBRIEFING 11:00
				CONTENT 2 (cont) - Conflict of interest; Data management	CONTENT 3 (cont) - Mentor and trainee responsibilities			LUNCH 12:00
		PRAYER 12:30						
LUNCH 13:00		LUNCH 13:00						DEPARTURES
Training 14:00	14:00	PEDAGOGY 1 - How People Learn		SMALL GROUP WORK Writing learning objectives and assessments	SMALL GROUP WORK Assessments and activities	SMALL GROUP WORK Activities and allignment	GROUP PRESENTATIONS	
		CONTENT 1 - A responsible scientist (1) Autism and the MMR vaccine			GROUP SHARE SESSION	GROUP SHARE SESSION		
		PRAYER 16:30						
		COFFEE BREAK 16:30						
	17:00	LARGE GROUP WORK Selection of topics for small group work		SMALL GROUP WORK Writing learning objectives and assessments (cont) FACILITATORS MEET 18:00	SMALL GROUP WORK Implementation of feedback from previous session FACILITATORS MEET 18:00	SMALL GROUP WORK Implementation of feedback from previous session FACILITATORS MEET 18:00	GROUP PRESENTATIONS	
WELCOME DINNER Introductions and administrative details 19:30	19:00	DINNER		DINNER	DINNER (on your own)	DINNER	FAREWELL DINNER	

FIGURE 4-1 Schedule at a Glance: Thematic and chronological structure of the Institute.

of work and group cohesion in some cases.

- *Allowing time for reflection and thought.* Silence among participants in a classroom or a group situation can appear to indicate lack of progress or disconnect between the facilitator and participants. However, well-used silence can help group members clarify their thinking and, sometimes, modify their positions about a contentious issue particularly prompted by group discussions. The effective facilitator builds silent periods into group sessions and tells participants the purposes for such periods.
- *Taking care to avoid becoming a participant.* When there are lulls in conversation or lapses of progress, it sometimes feels easier for facilitators to assume some of the roles and responsibilities of participants. Facilitators need to provide guidance and structure without taking over the group's agenda or its distribution of work. Indeed, if the group is progressing well in meeting its goals and plan of work, facilitators also need to recognize when to leave the group on its own. An effective group will need its facilitator less and less as the Institute progresses.

A number of publications are available to assist those who wish to replicate this type of facilitator training in their home institutions as part of a "train-the-trainer" program (see, for example, Branchaw et al., 2010:257-260; Pfund et al., 2012).

THE INSTITUTE ITSELF

Pedagogy Sessions

Throughout the week-long Institute, presenters who have been involved with the National Academies Summer Institutes for Undergraduate Biology Education (e.g., Pfund et al., 2009; Handelsman et al., 2006; Labov and Young, 2013) introduced key topics for effective, evidence-based teaching practices in three sessions: How People Learn, Assessment, and Active Learning. These sessions provided a framework for helping Institute participants transition from what cognitive science tells us about how people learn to practical applications for development of instructional material for the classroom and measurement of students' learning gains. A number of examples, resources, and references related to active learning are provided in Appendix D.

As in all sessions at the Institute, those making presentations and those facilitating discussions actively engaged participants in what was being taught and gave them practice with active learning and reflection. These themes were modeled not only in the pedagogy sessions but also in each of the content sessions so that participants could immediately apply the skills being taught during both types of sessions. This format also helped participants to better incorporate these concepts and skills into the modules they were creating, providing an environment where they could be both iterative and reflective about their learning (this kind of self-analysis of one's learning is termed "metacognition"; NRC, 2000).

Session 1: How People Learn

The first session, *How People Learn*, introduced participants to the essential findings from meta-analyses of the cognitive science literature, providing a rationale for why faculty should view teaching science differently than traditional norms and practices (NRC, 2012a). The goals of the session were to provide participants with a pedagogical framework for creating their RCR modules and for improving their teaching. The session highlighted that all learners come to the learning process with life experiences and preconceptions that often can lead to conceptual barriers to learning scientific concepts that are frequently non- or even counterintuitive, and that for any discipline we need to help the learner develop metacognitive skills and a conceptual framework for organizing information (content) and putting it in the context of other information.

Given that life experiences impact how people approach learning, it is important to view learners based on their worlds, rather than those of the instructors. For this reason, presenters addressed how college students (undergraduates and graduate) today are different from when most of the participants were in college. For example, the world is more globalized and information comes almost instantaneously from the Internet and through cell phones, which are relatively new modes of communication. In the United States, the so-called Millennial Generation faces different challenges from those of the Baby Boomers and Generation Xers, and these differences matter when it comes to their approaches to learning. Therefore, faculty should be aware of these differences and provide learning experiences that are well suited for these students. For example, spending time in class to provide students with primary content information is less necessary now than in the past since information is readily available anytime and anywhere. A larger issue is helping students make sense of this information, connecting it to other kinds of information and concepts, recognizing and addressing naïve or incorrect conceptions that they may have developed about some subject matter due to personal experience or being taught or learning a concept incorrectly, and helping them learn what information is bona fide and what is not, given that much less information is now vetted through trusted sources than previously.[43]

Instructors can also help students become more effective learners through reflection on their own learning and development of a conceptual framework for science (NRC, 2000). Providing learners with opportunities to review their learning progress is an important aspect of the learning process. It is also critical to help learners develop a conceptual framework, particularly through the practice of science, so they can more readily incorporate new content. Likewise, throughout the Institute the facilitator team challenged participants' own misconceptions, helped them reflect on their learning, and provided a pedagogical framework for developing modules.

The tenets of scientific teaching are that teaching science should be done with the same rigor, creativity, and general methodological approaches that one would apply to research, including the process of discovery (Handelsman, 2004). For example, when undertaking new research, scientists always search the literature to determine what is known about the subject, the methodologies used to investigate it, and how they can build on that body of knowledge. They develop hypotheses and design experiments to test them frequently. When a particular approach proves untenable, they redesign both the questions and approaches to addressing them. They share their data with other scientists both informally and through peer-reviewed papers. When employing scientific teaching, similar procedures would be used to design courses, teaching laboratories, and field experiences for students.

To help participants understand how to apply scientific teaching to their own classrooms, they were provided with a structure

[43] In some of the early learning literature (e.g., NRC, 2000), ideas that are incomplete or incorrect were referred to as "preconceptions" and "misconceptions," respectively.

for developing educational materials and learning experiences—"backward design"—that has been subjected to significant research to determine its efficacy (Wiggins and McTighe, 1998). Most traditional forms of teaching start by the instructor first designing the syllabus, selecting the text, and creating teaching materials, followed by construction of assessments. Although this approach may seem reasonable and is currently widely used by postsecondary faculty, it is mostly instructor-centered as the learning objectives and assessment of those objectives (exam questions) were designed after the teaching was complete. A more student-centered approach is to clearly state all measurable learning objectives and write associated assessments prior to instruction so that instruction is based on that template, keeping learners at the front and center of course development and the teaching process.

The concepts promoted during this session laid the foundation for how the Institute participants would develop their modules each day, and illustrated how they could change their teaching practices on their own campuses.

Session 2: Assessment

The second session, *Assessment*, introduced participants to new ways of thinking about assessment and how it can be used to improve learning as well as to measure learning gains. Presenters began this session with a discussion of the differences between summative and formative assessments and how these fit into theframework of backward design. Throughout the session the presenters emphasized the importance of articulating clear, measurable learning objectives for both guiding teaching and material development and enhancing learning. Hands-on activities allowed participants to practice writing measurable learning objectives at different levels—what are referred to in the research literature as *lower-order cognitive skills* (LOCS) and *higher-order cognitive skills* (HOCS) (see Figure 4-2; Crowe et al., 2008; Zoller, 1993). The lessons learned in this session helped participants to reflect on the use of assessment in their teaching and were immediately applied to the modules they were creating.

The main difference between summative and formative assessments is that summative assessment is the endpoint of measuring learning and formative assessment is measurement of learning throughout the learning process (Handelsman et al., 2007). Examinations (summative assessments) are the products of learning whereas assessments during learning (formative assessments) can help guide the instructor and learners to change their practices and strategies for teaching and learning, respectively. Summative assessments can be given in many forms (exams, written papers, final presentations or some other form of work). Formative assessments can include techniques such as short quizzes at the beginning or the end of class sessions for which students receive a few or no points, or questions during class where students can state their answers using flash cards that they hold up or through the use of electronic response systems (also known as "clickers").

Because summative assessments are high- stakes for the learner, they drive learning and therefore can be powerful learning tools. The first part of the session stressed how educational materials, including the modules that participants would develop during the Institute, should be designed in ways that best guide the learner through the learning process, and that the instructor's intentions and expectations for assessments should be stated at the outset.

A clear statement of what learners should know or be able to do (learning objectives) before they are taught the material can help guide them in the learning process, particularly

Cognitive Level	Bloom Level A Simple Phrase to Guide Categorization
HOC	**Evaluate** "Defend or judge a concept or idea"
	Synthesize "Create something new"
	Analyze "Distinguish parts and make inferences"
LOC/HOC	**Apply** "Use information or concepts in new ways"
LOC	**Comprehend** "Explain information or concepts"
	Know "Recall information"

FIGURE 4-2 Bloom's Taxonomy and Cognitive Levels. HOC, higher-order cognitive skills; LOC, lower-order cognitive skills. The original levels of cognition proposed in 1956 have been modified by others. For example, Overbough and Schultz (*http://ww2.odu.edu/educ/roverbau/Bloom/blooms_taxonomy.htm*, date unknown) have proposed the following descriptors ranging from the lowest to highest levels: Remembering, Understanding, Applying, Analyzing, Evaluating, Creating. Additional sources of information about these modifications to Bloom's Taxonomy are available through links on the website by Overbough and Schultz. SOURCE: Created by the committee.

if they understand that the learning objectives are tied to summative assessments. Instructors can use the framework of backward design not only to align the content of their formative and summative assessments but also to adjust the cognitive levels at which the learner is engaged during learning and testing. Around any content area, the instructor's learning objectives can be directly tied to, and aligned with, both summative and formative assessments. To this end, participants engaged in a series of activities that allowed them to practice aligning summative and formative assessments so that they could better understand the relationships between these two important aspects of learning.

An important aspect of aligning summative and formative assessments is that both the content and the cognitive level at which the learner must work should be taken into consideration. Most instructors find it fairly easy to align what they teach with what will be on the test, but a more difficult task is aligning the challenge level of the content practiced and tested. Bloom's *Taxonomy of Education Objectives Handbook I: Cognitive Domain* is a classification system to distinguish six categories (see Figure 4-2) or levels of human cognition and has been effectively used for over 50 years to develop curricula (Bloom, 1956). As mentioned above, one of the most useful distinctions lies not in the differences among the six categories but rather in the difference between categories that require higher-order cognitive (HOC) and lower-order cognitive (LOC) skills (Zoller, 1993). Simplifying the taxonomy into two groups helps one to quickly assess how challenging the learning objectives and assessments will be to the learner (Crowe et al., 2008). After introducing this concept, participants learned to categorize questions and activities based on Bloom's Taxonomy.

During this session of the Institute, participants focused on the importance of assessment and how different forms of assessment are related to teaching and learning. The presentation and discussion centered on backward design as a way to align content and Bloom's Taxonomy to help evaluate how content and cognitive levels could be incorporated into the modules that participants would develop during the small group work in the afternoons and evenings (see below).

Session 3: Active Learning

The *Active Learning* session emphasized how faculty can transition from an instructor-centered to learner-centered approaches using a repertoire of techniques for engaging learners (see Table 3-1). Active learners take responsibility for their learning by participating in problem solving, group work, or related activities that engage them in the learning process and help them construct their knowledge. With a "toolbox" of active learning techniques that were provided during these sessions, Institute participants were assisted in developing RCR modules that incorporate evidence-based best practices in pedagogy that they would use themselves and disseminate to their colleagues.

An important aspect of this session was to help participants realize that all learners, including themselves, have preconceptions or misconceptions about content and that those misconceptions need to be addressed for successful learning.[44] An effective method for teaching this concept was to model how formative assessment and active learning can uncover common misconceptions in science. Therefore, the presenters designed activities that would engage participants in the ways that students are often challenged when learning new scientific concepts. Participants were given a problem, which they first considered independently, followed by group discussion. The process was repeated until the larger group was able to correctly solve the problem. Through this method, participants learned the importance of group work in problem solving, i.e., that learners can often help themselves and others with whom they interact to learn independently from their instructors, but also how difficult it can be for some to overcome their misconceptions. By illustrating how active learning not only engages the learner but also helps creates cognitive dissonance for those with misconceptions, participants were shown the value of active learning and brought this understanding to their module development.

During this session, presenters also shared the wealth of data from the literature on science, technology, engineering, and mathematics (STEM) education that supports the effectiveness of these methods and demonstrates their use for enhancing learning. Scientists are receptive to changing or refining their views based on evidence. An increasing amount of data from rigorous studies shows that active learning helps more students succeed academically in the science disciplines. Presentation and discussion of this evidence helped foster the participants' understanding that these methods work and that faculty should not approach teaching and learning based on traditional norms and practices but rather through actively engaging learners throughout the learning process, assisting them in developing lifelong learning and collaborative skills.

An important part of this pedagogy session was to model the many ways in which active learning can be implemented. Throughout the content and pedagogy sessions the Institute engaged participants in a variety of ways and made learning objectives for each session transparent so that the participants could

[44] There is an emerging literature on misconceptions in many fields of science. For example, for misconceptions about various aspects of biology, see Coley and Tanner (2012).

practice aligning learning objectives with formative and summative assessments in multiple ways. Presenters were explicit and reflective about the methods used, and throughout the week this transparency, openness, and willingness to engage participants with their own learning helped them better understand how to use a wide variety of active learning techniques in their own teaching. This session also summarized the third step in using backward design in that participants then added to their modules the ways in which they would engage the learning in connecting with the content the module was designed to deliver.

Content Sessions

Following the planning meeting in Trieste, the content group discussed how to organize the workshop around the responsible conduct of research topics suggested by the National Institutes of Health (NIH, 2009) and expand them to include a discussion of research with dual use potential. That discussion led to the three themes that were the focus of the content sessions of the Institute. Those themes dealt with one's obligations to be responsible scientists in one's daily professional life, to conduct research responsibly, and to be a member of a community of responsible scientists. The three themes and included topics are presented below. Most topics were covered under more than one theme. The cases were constructed or adapted around themes with universal recognition that affect scientists in similar ways regardless of country of residence. Specific cases, also summarized in the section below, were chosen to emphasize one or more of the topics in each theme. Background readings for the cases were available to the participants on a password-protected website before the Institute. Importantly, committee members with expertise in the content to be discussed during the Institute worked closely with the committee with expertise in science education and pedagogy for weeks prior to the Institute to plan active learning pedagogies for integration into the actual Institute sessions.

Theme 1: The development of professionalism in science. Discussion and analysis focused on the development of professionalism and the role of government regulations and institutional policies. The session introduced elements of research misconduct using the case study *Autism and the MMR vaccine.*

Theme 2: Conducting research responsibly. Discussion and analysis included working with and protecting human subjects, humane and ethical care and use of laboratory animals, conflict of interest, and data management. The case studies were *Introducing viruses in the field* and *The Guatemala syphilis studies.*

Theme 3: Being part of the responsible scientific community. Discussion began with the topics of collaborative research, authorship and publication, and peer review using the case studies *The Darsee affair* and *Who is an author?* Additional discussion focused on mentor and trainee responsibilities; research with dual use potential using the slide presentation *Potential threats from biotechnology and life sciences: What is dual use research?*; and biosafety concerns in research through the case *Studies in H5N1 influenza virulence.*

Pathways and Inspirations: A Conversation with Institute Participants About Being a Scientist

The first day of the Institute began with a conversation about the professional commitments that had brought both facilitators[45] and participants to the Institute and the meanings that theyhave found in our respective pathways in science. Three of the Institute faculty, Nancy Connell, Alastair Hay, and Elizabeth Heitman, recounted personal stories of their careers, reflecting on the struggles, successes, and surprises they have encountered. Participants were then asked to pair with a new colleague to learn how the other had become interested in science, the context and scope of his or her current work, and the particular points along the way at which he or she found professional and personal meaning. Each participant then stood and publicly introduced the new colleague to the larger group, focusing on the information or experiences that seemed most important or characteristic of that person.

Three common themes emerged from the introductions: the sense of calling that many scientists experienced first as students that often continued throughout their careers; the experience that unforeseeable events had often been crucial to participants' research focus and career trajectories; and the ability of strangers from different institutions and different disciplines to find common ground in their stories of science. This exercise, in addition to initiating participants into the processes of active learning, demonstrated the broader meaning of the concept of "scientific community" on which the Institute was built.

Case 1: Autism and the MMR vaccine[46]

Consideration of Theme 1, Development of Professionalism in Science, was facilitated by a discussion and analysis of the controversy surrounding the purported causative relationship between autism and the MMR vaccine in children. In 1998, Andrew Wakefield and colleagues published a paper in the *Lancet* titled "Ileal-lymphoid-nodular hyperplasia, non-specific colitis, and pervasive developmental disorder in children." His hypothesis was that the measles, mumps, and rubella (MMR) vaccine causes a series of events that lead to the development of autism. In support of his hypothesis, Wakefield described 12 children with neurodevelopmental delay (8 with autism). All of these children were reported to have had gastrointestinal complaints and to develop autism within 1 month of receiving MMR. But there were a number of critical flaws in the experimental design and conduct of the reported study and the paper was eventually retracted. However, equally serious issues were those of research fraud, unethical treatment of vulnerable children, and conflicts of interest.

This case ultimately resulted in greatly reduced numbers of children receiving life-saving vaccines and untold anxiety for parents making decisions about their children's health care. By falsely linking autism to vaccines, Wakefield created an international crisis in preventive medicine. This case was chosen because it demonstrates a number of important concepts and principles regarding professionalism, including the

- importance of data selection and presentation to research integrity,
- importance of disclosing financial conflicts of interest on research ethics,
- responsibilities of coauthors for study design and interpretation,
- appropriateness of presenting research findings in press conferences,
- ethical concerns that can arise when conducting research with children,

[45] The terms facilitator and institute faculty are used interchangeably in this report.

[46] Original case developed for the Institute. The background readings for this case were Wakefield et al. (1998) and Horton (2004). Additional readings include Deer (2011a,b,c) and Pilonis (2007).

- importance of revealing study sponsors to participants in human trials and institutional review boards, and
- potential impact of research impropriety.

The active learning strategy chosen for this session was large group discussion of the case, where the facilitator presented the case and then encouraged the participants to contribute to the discussion of the topic. This technique was appropriate for the first session of the workshop since many of the participants were somewhat familiar with the circumstances surrounding the case and how to approach case studies as a learning tool.

Case 2: Introducing viruses in the field[47]
Theme 2 was Conducting Research Responsibly, and the session focused on research with animals and human subjects. Specifically, the facilitators wanted to address scientists' responsibility in protecting research subjects (both animals and humans) as well as the communities in which the experiments are carried out. The themes of experimental safety and animal protection were highlighted in this approach.

The first case study concerned a hypothetical proposal to test a live vaccine on a population of chimpanzees living on an island. The case described the quandary of a young investigator whose expertise in primate biology earned her a position on the animal protection committee of her institution. She was charged with the review of a proposal to test an altered live virus vaccine for hepatitis C (HCV) using a free-ranging chimpanzee colony. This colony has been established for behavioral research studies 20 years earlier on an island near Puerto Rico. The colony receives daily food supplementation by boat. The research plan is to inject one of the dominant males with wild-type HCV and to vaccinate half of the remaining animals, both males and females, with a live recombinant virus vaccine. All chimpanzees are to be monitored for the development of viremia, immune responses to the virus, altered liver function, and chronic infection. An additional protocol will use the animals that become infected for a clinical trial of new chemotherapeutic agents. Chimpanzees were selected because, like many humans, they have multiple sexual partners and are susceptible to the virus. Since HCV naturally infects only humans and chimpanzees, the research group felt that it was necessary to get a definitive answer under field conditions before introducing live recombinant viruses into uninfected human populations.

The following questions were proposed for discussion:

- What are some of the troublesome issues associated with this set of experiments?
- What are the specific concerns about administering a live recombinant virus to humans as a vaccine?
- Is it ethically appropriate to intentionally transmit a human virus in a setting that is not fully controlled?
- If it was decided that the study could not be carried out in chimpanzees, how might it be designed instead for human subjects?

The active learning technique used during this session was similar to a "jigsaw" wherein individuals in a group reach consensus about a position or gain expertise about a topic, after which new groups are formed so that one person from each original group informs the others in

[47] Case and questions adapted from *Introducing Viruses in the Field* (National Academy of Engineering Online Ethics Center for Engineering 9/10/2006; *www.onlineethics.org/Resources/Cases/HIVan.aspx*. Additional background reading is available at *http://ori.hhs.gov/education/products/ncstate/models.htm*.

the new group. The facilitators used a similar approach in that groups discussed one of the four questions above, came to consensus, and then each group contributed to the larger discussion. The general consensus of the participants was that the experiments should not be allowed to go forward, citing safety and concerns about release of infectious virus. There was a lively discussion of the challenges of developing animal models for human disease. The participants recognized and elaborated on the responsibilities of researchers for environmental and community safety.

Case 3: The Guatemala syphilis studies[48]
Continuing the theme of Conducting Research Responsibly, this case addressed the ethical standards of research with human subjects and the harm that research on infectious diseases may cause when research participants' interests are made secondary to scientific goals. The case of U.S. Public Health Service (USPHS) research on "normal exposure" to syphilis was chosen because it reflects many practical and ethical challenges in today's infectious disease research. It also demonstrates the development of comprehensive ethical standards for epidemiologic research and provides a stark example on how professional dedication to an important scientific goal can blind researchers to ethical considerations relevant to their work.

The case took place in the mid-1940s, when syphilis still caused widespread death and disability. During this period, the USPHS explored various uses of penicillin in preventing and treating syphilis in populations where it was reported to be endemic. In one such series of studies, internationally known syphilis expert Dr. John Cutler led U.S. and Guatemalan researchers in experiments designed to test penicillin as a prophylactic against "normal exposure" to syphilis. Between 1946 and 1948, Dr. Cutler's group paid syphilis-infected Guatemalan sex workers to have sex with uninfected prison inmates to measure rates of transmission. Additionally, some uninfected women had syphilis inoculum placed on their cervix before they had sex with uninfected prisoners. Later, researchers conducted an inoculation study in a Guatemalan institute for the mentally ill. Participants who tested positive were treated with a presumed curative dose of penicillin, but few were told that they were being given live doses of syphilis as part of a study. Researchers acknowledged privately that this work was ethically controversial, but many were eager "to study syphilis from the standpoint of pure science."

Institute participants used the "think, pair, share" method to examine several conceptual and practical issues, including:

- how the scientific method shapes the risks to which research participants may be exposed,
- how the perceived threat posed by an infectious disease affects the assessment of risk and benefit associated with related research,
- whether intentional exposure to disease might be acceptable in research,
- the perceived advantages of undertaking infectious disease research in developing countries, and
- the perceived advantages and risks of international collaboration for researchers from developing countries with those from more scientifically developed nations.

This case was unfamiliar to most Institute participants. From the background reading and

---

[48] Original study developed for the Institute. The background readings for this case were Reverby (2011), WMA (2008), and the report of the Presidential Commission for the Study of Bioethical Issues (2011). (Although the full report was too long to be included, the preface was included and the report discussed.)

discussion, participants developed knowledge of the practical challenges of conducting prevention research on serious infectious diseases in ways consistent with today's ethical standards for international infectious disease research. The group agreed that the ethical standards of today generally protect the welfare of human participants and safeguard the quality of the research results, and that this balance demands the researcher's responsible conduct of science on many levels.

Case 4: The case against John Darsee[49]

To conceptualize the topics of collaborative research, authorship and publication, and peer review, the case of Dr. John Darsee was used to introduce Theme 3. Darsee was highly regarded as a student and medical researcher throughout his undergraduate and postgraduate training. At Harvard University, he worked as a research fellow in the Cardiac Research Laboratory headed by Dr. Eugene Braunwald. His special area of research concerned the testing of heart drugs on dogs. In less than two years at Harvard he was first author on seven publications in very good scientific journals.

In 1981, colleagues in the Cardiac Research Laboratory observed Darsee mislabeling data recordings from an experiment he was performing. Over the next several months, it became clear that Darsee had been fabricating or falsifying data for years, possibly back to his undergraduate days. The consequences were profound, not just for Darsee but for the members of the laboratories where he had conducted his investigations, his mentors and supervisors, the coauthors on his published manuscripts, the institutions where he had worked, the scientists who had relied on the veracity of his research reports to shape the direction of their own research, and the patients whose treatment may have been influenced by his publications.

Some positive things have come from the Darsee case. In addition to alerting scientists to the need for providing closer supervision of trainees and taking authorship responsibilities more seriously, the Darsee incident contributed to the development of guidelines and standards concerning research misconduct by the USPHS, U.S. National Institutes of Health, U.S. National Science Foundation, medical associations and institutes, universities, and medical schools.

This case was chosen because it demonstrates a number of important concepts and principles related to responsibilities of individual scientists as members of a larger responsible scientific community, including:

- ethical concerns that can arise when conducting collaborative research,
- the essential responsibilities and professional relationships of mentors and trainees,
- responsibilities of authorship, and
- flaws in the peer review system.

To engage participants, the facilitator presented the case and then each group discussed the case. A spokesperson for each group presented the group's consensus and this was followed by a larger group discussion. This format was appropriate for this case since many of the participants are academicians and the issue of authorship responsibilities was particularly relevant to them. There were individuals in each group at various stages of academic development, so the different perspectives could be shared.

[49] Case and questions adapted from "Case Study 1: Overly Ambitious Researchers - Fabricating Data" National Academy of Engineering Online Ethics Center for Engineering 7/20/2006; *www.onlineethics.org/Education/precollege/scienceclass/sectone/chapt4/cs1.aspx*. The background reading was Kochan and Budd (1992). An additional reference is NAE (2007).

Case 5: Who is an author?[50]

Continuing the theme of Being Part of the Responsible Scientific Community, this case addressed the common and often contentious issue of the qualifications for authorship and the collegial responsibilities that come with research publication. During the discussion, participants put themselves in the role of a young investigator who is preparing to submit an article based on a collaborative research project.

The protagonist is in the sixth month of a two-year research fellowship at an academic medical center, with hopes of joining the faculty. One part of the fellow's work is to continue a line of research originally started by a junior faculty member who left the university to have a baby. The former faculty member is now a physician in private practice. Before she left, she designed the original protocol and collected blood samples and data on 40 patients. With the help of a former biostatistics professor, the research fellow revised the methodology and got Institutional Review Board approval for a new protocol. The research fellow worked with a resident physician and a nurse, collected blood samples and medical histories from an additional 145 patients, and then analyzed the samples and data from all of the samples with the help of an undergraduate microbiology student research assistant. The research fellow wrote up the manuscript and made two data tables with the biostatistics professor. The department head provided edits and helpful suggestions, but also emailed the manuscript to the former faculty member asking for her insights. She replied to the research fellow, asking to be the second author because she wanted to return to academic medicine.

Using the modified jigsaw method of case discussion, as described above and in Table 3-1 each of the five groups discussed one of the following and presented their position to the larger group:

- the arguments for including the former faculty member as an author and what other actions, if any, she would need to undertake with this manuscript to qualify,
- the arguments against including the former faculty member as an author, and what to tell the department head,
- whether to include others (the biostatistician, resident physician, nurse, undergraduate microbiology student, department head) as authors, and
- the order in which to list the named authors and the qualifications for each position.

Questions about the qualifications for authorship lead to some of the most significant disputes in science, and such conflict affects many academic communities. In the background reading and discussion, participants examined the importance of assigning authorship on a manuscript at the beginning of a collaborative project. Using the criteria for authorship from the International Committee of Medical Journal Editors (ICMJE) Uniform Requirements for Manuscripts, participants developed strategies for preventing confusion or conflict over authorship by agreeing on criteria and group expectations early in a project.

Most participants at the Institute were published authors and all agreed that publication—particularly in English-language journals—was important to their careers. A number of participants knew of the ICMJE's Uniform Requirements, and many found it useful to examine how they applied in specific circumstances. Several individuals noted that they would use these standards in their own work, particularly in discussion with superiors and trainees.

---

[50] Original case developed for the Institute. The background readings were Albert and Wager (2003), ICMJE (2010), and WAME (2013).

Research with Dual Use Potential

Great achievements in molecular biology and genetics have produced advances in science that have revolutionized the practice of medicine. The very technologies that fueled these benefits to society, however, pose a potential risk as well—the possibility that these technologies could also be used to create the next generation of biological weapons. Under Theme 3, Being Part of the Responsible Scientific Community, discussion of potential threats from biotechnology and life sciences was facilitated by a PowerPoint presentation. For the purposes of this discussion, dual use was defined as "research that, based on current understanding, can be reasonably anticipated to provide knowledge, products, or technologies that could be directly misapplied by others to pose a threat to public health, agriculture, plants, animals, the environment, or materiel."[51] A number of types of risk from dual-use research were discussed, including:

- technologies that deliver beneficial drugs to the body could be used for weaponizing biological agents,
- research could have unintended consequences,
- dangerous agents could be released accidentally from the lab through infected personnel or other means (e.g., faulty exhaust systems),
- research results and methods can be published in easily accessible journals and on the Internet,
- knowledge or techniques could help to create "novel" pathogens with unique properties or create entirely new classes of threat agents, and
- dangerous agents could be stolen or diverted for nonpeaceful purposes.

Several examples were discussed, including that any medical advance that improves the ease of engineering, handling, or delivery of treatment has the potential to be applied by those wishing to do harm and can be considered "dual use," that each year hundreds of articles on dual use research are published, making them accessible to any member of the research community, that thousands of pieces of scientific equipment are purchased on the Internet without oversight or regulation, and that this openness creates the risk that available information, reagents, or equipment might be used to create new or more dangerous biological weapons.

The presenter used the Socratic method by posing specific questions that elicited discussion.

- Does a select agent list make us more or less safe?
- What steps can be taken to ensure that resources/equipment/knowledge are not used inappropriately?
- Is it a question of "Who could" or "Who *would*"?

The active learning strategy used for this session was a presentation followed by group discussion, where the facilitator reviewed the key issues, each group discussed the issues among themselves, and then a spokesperson for each group summarized the group"s consensus. This format was appropriate for this case since the experience level of the participants varied so widely.

[51] This is the definition adopted by the U.S. National Science Advisory Board for Biosecurity (NSABB, 2007). See Box 1-1 in Chapter 1.

Case 6: Studies in H5N1 influenza virulence[52]

Following on the PowerPoint-led discussion on experiments with dual use potential, this case focused on two sets of highly controversial experiments carried out with influenza virus H5N1, an avian influenza strain that has shown alarming morbidity and mortality in the limited number (fewer than700) of human infections that have occurred since its identification in 1997 in Hong Kong. To discuss the experiments, it was important to establish first that the participants' knowledge base was at a sufficient and similar level. Therefore, the session began with each group working together to standardize their backgrounds in influenza biology. Afterward, a brief outline was presented to describe the series of events that led to the crisis in H5N1 research and subsequent moratorium on continued experimentation, imposed in January 2012.

The controversy began when influenza researchers announced in September 2011 at a flu conference in Malta that they had created mutant forms of the H5N1 influenza virus that were transmissible between ferrets. The two research labs involved submitted manuscripts to the journals *Science* and *Nature*. The National Science Advisory Board for Biosecurity (NSABB) of the United States was asked to review the two manuscripts. The NSABB recommended that the manuscripts be revised and published with redacted details on the specific mutations and with additional discussion of public health values of the work, as well as a description of increased safety and security research practices. A group of leading flu researchers declared a moratorium on the type of research that had caused the controversy pending international discussion. The World Health Organization (WHO) held an international meeting in mid-February 2012 with 22 scientists and public health experts who concluded that the work should be published in full after the moratorium. At the end of March of the same year the NSABB voted that the two papers should be published in their entirety after reviewing revised manuscripts and receiving additional information. The papers were published in the summer of 2012, and the flu researchers ended their moratorium in January 2013.

Before the case was presented, each group was asked to prepare a poster about the H5N1 replication cycle and pathology so that those less familiar with the content were informed. After this exercise, the case was presented and the participants were asked to work in groups to address the following questions representing the key points that this case study presents:

- Why are these experiments deemed to be dual use research of concern? Summarize the experiments and discuss the possible nefarious uses of the information that might be gained from these experiments.
- Should scientists perform the research or not?
- The uproar surrounding the two recent H5N1 studies spotlights the issue of whether or not research on potentially dangerous lab-generated pathogens should have been conducted in the first place. What are the benefits and the risks? Do the benefits outweigh the risks?
- Should the results be published or not?
- Should this kind of work be regulated?
- A global issue that stems from the ongoing H5N1 debate is how to regulate such research. Who should be in charge of granting approval for potentially dangerous studies? At what biosafety level should they be conducted? Who should have access to

[52] Original case developed for the Institute. The background readings for this case were Berg et al. (1975), Enserink and Malakoff (2012), Maher (2012), Morris (2012), and *Nature* (2012).

the full results? How should all of this be organized and monitored?

This session also used role playing as the active learning technique. Each group was assigned a single role to play in the unfolding crisis, which led to animated debate among the groups. The group assignments were (1) the authors, who support and justify the research approach, (2) the NSABB, which initially opposed the publication of the experimental details, (3) WHO, which recommended the publication of the experimental details, (4) the public, who were frightened and skeptical, and (5) the media, which tend to use inflammatory language to promote a story. Each group discussed its position and then defended its position to the larger group while staying in its assigned role.

The learning objectives for this case touched on many of the issues covered in other sessions, such as the responsibility of researchers for the safety of the environment and community in the design of experiments. The focus of the case on dual use allowed participants to define "dual use research of concern" in the context of a real-life event and to explore the ethical and regulatory issues related to the experiments. Finally, the participants were asked to identify the issues surrounding the debate over publication of experimental details that might lead to creation of dangerous material.

Small Group Work

As previously discussed, participants were assigned to five groups at the beginning of the Institute, so that whenever possible each group included participants from all countries without overlap of participants from the same team or institution. Groups received brief descriptions of their assigned topic as well as a number of questions to address (see below). Two facilitators were assigned to each group, while two more functioned as "floaters," providing their pedagogy and content-related expertise to all the groups. Each group was tasked to work independently during the afternoons to develop teaching modules around their assigned topic using active learning and appropriate assessment techniques; the groups' presentations are summarized below. Each group had opportunities throughout the week to share its ideas and presentations with other working groups and was required to make a formal presentation to the entire Institute at the end of the week. By the end of the week, each group had developed a peer-reviewed, teachable unit on some aspect of responsible conduct of science and had learned how to implement scientific teaching and mentoring workshops on their own campuses.

Topics for the Small Working Groups

Research Misconduct

Misconduct is defined as fabrication, falsification, or plagiarism in proposing, performing, or reviewing research, or in reporting research results. Fabrication is making up data or results and recording or reporting them. Falsification is manipulating research materials, equipment, or processes, or changing or omitting data or results such that the research is not accurately represented in the research record. Plagiarism is the appropriation of another person's ideas, processes, results, or words without giving appropriate credit.[53] Research misconduct also encompasses the failure to comply with legal requirements for protecting researchers, human and animal subjects, and the public. It is important to

[53] See *http://ori.dhhs.gov/definition-misconduct.*

understand that research misconduct is not an honest mistake in reasoning, differences of opinion, disagreeing with recognized authorities, misinterpreting results, an error in planning or carrying out an experiment, or an oversight in attribution (ibid).

Questions for Discussion

- Should other practices besides fabrication, falsification, and plagiarism be considered research misconduct?
- Is it fair to use "significant departure from accepted practices" to make judgments about a researcher's behavior?
- Should researchers report misconduct if they are concerned that doing so could adversely impact their careers?
- What evidence is needed to demonstrate that a researcher committed misconduct intentionally, knowingly, or recklessly?
- What are appropriate penalties for different types of misconduct?

Group Presentation

Given that the groups were asked to plan how to teach others, most of the presentation centered around slides that defined the goals, objectives, teaching sessions, and assessment procedures to be undertaken. The goal set out by the group was to encourage those being taught to think critically about types of misconduct and during this process to discuss categories of misdeeds, causes and consequences, and reporting strategies for communicating to others. The presentation outlined a range of assessment procedures for summative measures. These were aligned with various active learning approaches including discussion, case studies, jigsaw approaches, and encouraging participants to voice an opinion on questions raised during teaching sessions, either by using clickers or simply raising a hand. The discussions envisaged by the group would cover research misconduct and what might prompt this, as well as what would prompt individuals to plagiarize, fabricate, or falsify data. At the end of their presentation the group posed a question to the audience, inviting them to consider ten major reasons for misconduct and then discuss them.

Responsible Authorship

Writing research papers is an essential activity for a scientific career and for the scientific process. Articles are important for academic recognition and authors have a responsibility to publish their results to further the scientific enterprise. But the scientific process and scientific publication have changed significantly over the years. Research has become more competitive, complex, and multidisciplinary, with collaborations among senior scientists, clinicians, undergraduate and graduate students, technicians, postdoctoral fellows, medical students and residents, statisticians, and other professionals in both national and international contexts. Each brings different expectations and experiences to issues such as who should be included as authors in a paper for publication and the value of their respective contributions. Good scientific practices include discussions before, during, and after the research process to ensure that the allocation of authorship is ethically determined, along with sound study design and attention to the protection of human subjects and the ethical use of animals. Individual journals have guidelines for authors, and the ICMJE meets regularly to update the Uniform Requirements for Manuscripts Submitted to Biomedical Journals (*www.icmje.org/index.html*) to address emerging issues.

Questions for Discussion

- What does it mean to be an author of a scientific paper?

- What are the different positions in a list of authors and what does each position signify? Why is the order of authors important?
- How would you define "prior publication" and why is this important when considering authorship?
- What is the difference between acknowledgment and listing as an author?
- Who takes responsibility for submission and follow-up of revisions, etc.?
- What are the problems associated with using the same data in multiple publications?
- Should all authors be responsible for all of the information in the paper?
- What are some of the problems that might ensue from publishing results early, before complete confirmation?

Group Presentation
The presentation began with a cartoon depicting a too familiar situation: a young researcher being informed that his chances of publication would be augmented by having his laboratory chief as first author of the paper. The goal of the group was to ensure that those being taught would gain a clear idea of what responsible authorship meant, why it was important to publish, and to understand what would motivate individuals to do so. Slides were used to inform the workshop participants how the group would approach its teaching. The group presented a case and asked the workshop audience to discuss whether it was appropriate that a certain individual be an author. The participants were then asked to use clickers to answers six questions about who might qualify as an author, ranging from those doing laboratory work to others providing statistical advice.

Collaborative Science
Collaborative science is the process of conducting research as a team of multiple individuals across laboratories, departments, institutions, and/or disciplines. Collaboration in life sciences research is increasingly more international in scope and partnerships are more and more diverse. While collaborations have been a common characteristic of almost all scientific inquiries for over 50 years, a number of problems can arise. Researchers have different styles of research, conferences, journals, language, ethics, standards, and schedules. Misunderstandings and conflicts caused by these differences can lead to undue stress on the group. The best way to anticipate these kinds of problems is to address potential conflicts before the work is begun or immediately as they arise.

Questions for Discussion
- What are the various kinds of collaborations about which you know?
- What factors drive the increase in collaborative and multidisciplinary research?
- What are the kinds of problems that collaborators face?
- What are some mechanisms that might prevent conflicts between and among collaborators?
- What are the essential elements of successful collaboration?
- How can institutions promote and support successful collaboration?

Group Presentation
Group 3 made use of visually arresting slides to provide some theoretical background on the meaning of collaboration before turning to the necessary goals, objectives, and teaching approaches. Flipcharts, clickers, handouts, and audience question and answer approaches were all used to convey the message. To illustrate collaboration further, the group presented a real-life case involving a number of countries researching the antimicrobial properties of essential oil from a plant found in the Mediterranean region, *Juniperus communis.*

Issues that the group identified as vital for discussion with others included the role of each individual in a collaborative project, the time frame, finances, potential conflicts of interest, and whether working with others provided any added value. Finally, the audience was asked to consider who ought to be first author and was invited to use clickers to address a range of options.

Mentor-Trainee Relationships and Responsibilities

Academic scientists traditionally have three interrelated and complementary roles: they conduct research, they teach students, and they provide service to society. Undergraduate-level teaching in science typically focuses on students' general knowledge and basic laboratory skills. Graduate-level teaching is focused on the deeper knowledge and complex abilities that trainees need to become independent researchers. Graduate programs typically assign each trainee an academic advisor and research supervisor to oversee their academic progress, but most successful young researchers can also point to one or more mentors. A mentor is typically a more senior researcher who takes special interest in guiding a trainee's development as a professional. The role of a mentor may vary with the discipline, institution, and type of research, as well as the personalities of the mentor and trainee. A trainee may also have different mentors in different areas of his or her work. Because faculty have a great deal of authority over trainees, these relationships also hold the potential for abuse. Problems can arise when faculty and trainees have different expectations of their roles and responsibilities, particularly in regards to workload and allocation of time, authorship credit, standards of productivity, and relationships with other faculty.

Questions for Discussion

- What are some of the qualities of a great teacher? How are these similar to and different from the qualities of a good mentor?
- Why might a successful researcher want to be a mentor to a student just entering science?
- What responsibilities do trainees have to their research supervisors?
- Some universities require advisors and trainees to create a written agreement about their future work together. How can such a document help or harm a mentoring relationship?
- How might a younger researcher serve as a mentor to an older scientist?
- Does a researcher ever stop needing a mentor?

Group Presentation

The slide presentation by Group 4 began with an arresting quote to make clear that the audience would not be able just to sit back and listen to speeches. Audience involvement was expected; clickers were used to collect responses. The group noted that the issues would present many challenges and opportunities for conflict and participants discussed some of the potential sources. Quality—that is, what was needed for good mentoring—and the responsibilities of mentors and trainees were recurring themes. The group proposed using the case of a doctoral student whose research program became vulnerable through the absence of a supervisor. The student persisted with the work after various discussions with others, but the final doctoral submission was rejected by the supervisor as inadequate. Trainees would be invited to explain what steps each participant in the student's program ought to have followed, including the student himself, the supervisor, the head of department, and the academic board.

Safe Laboratory Practices: Keeping the Community Safe

Laboratory safety is an essential feature of a responsible scientific enterprise. The development of vaccines and other prophylactic, diagnostic, and therapeutic interventions (e.g., antibiotics) for the treatment of infections requires increasingly complex experimental methods that pose complex risks. Infectious disease research typically requires the use of animals to model human disease. As the scope and amount of infectious disease research has expanded, there has been an increase in the risk of laboratory-acquired infections among research personnel. To protect their workers and the surrounding community, laboratory directors must incorporate good laboratory practices into their programs, and young scientists must be trained in laboratory safety. Global efforts to create a code of conduct for life scientists have tried to address the following issues: first, do no harm; second, ensure the safety of laboratory workers and the surrounding communities; third, incorporate the principles of the Biological Weapons Convention into daily practice.

Questions for Discussion

- What are the regulatory bodies in your country that oversee laboratory safety?
- How are students taught about laboratory safety and safe practices?
- What is the reporting structure in the event that you perceive unsafe practices?
- Is there protection for people who report unsafe practices?
- What kinds of laboratories should be used for dangerous experiments?
- What is the Biological Weapons Convention and how does it apply to biomedical research?

Group Presentation

Once again using slides, Group 5 began by explaining what safe laboratory practice is. This was followed by the goals for its teaching program. Teaching approaches would be very "hands-on," emphasizing rigorous attention to detail, and trainees would have to understand the purpose of specific containment procedures for organisms with different risk profiles. Periodic assessment was envisaged for students to test their knowledge and practice. The group introduced the case of a student who realizes that the virus s/he has extracted from cells is high-risk and categorized as a bioterrorism agent. Again using clickers, the audience was invited to choose an answer about what the student should do from a proffered list. As its parting display the group presented a small play in the form of a silent movie. With a facilitator narrating, a group member spilled an unknown liquid (in the form of bits of paper), which a laboratory worker discovered. The play illustrated the sequence of events to follow to warn others and then both contain and clean up the spill, including donning plastic shopping bags to simulate disposable laboratory boots and white ladies' gloves as their laboratory equivalent.

Facilitator Debriefing

Each day during the Institute the facilitators met—typically toward the end of the day and while the groups were working on their projects—to share their successes and challenges, compare approaches to facilitation that might be used by others, and agree on goals and work for the next day's group sessions. A final session, which took place after the Institute ended so that facilitators could reflect on the entire week of group work, is described in Chapter 6 as part of the evaluation process.

These sessions enabled the facilitators to identify issues that impeded the optimal function of the small groups, particularly in the initial stages of collaboration. For instance, a policy at NASI requires all members of a team from a college or university to work together in small groups to develop their teachable units. However, the dynamics in some groups at the Institute resulted in the facilitation team's decision to declare that no two people from the same institution could work together in the small groups. This action enabled all participants to take part in group discussions without concern for how they were perceived or the need to defer, especially in the case where teams were composed of a senior faculty member and more junior colleagues. Facilitators also used these meetings to plan for the next day, which allowed them sometimes to alter assignments in response to the dynamics of the small groups.

Toward the end of the Institute, one team of facilitators began designing the survey that was administered three weeks after the completion of the Institute. A separate team designed the Request for Applications (see Appendix G) that was disseminated to all participants. Through that mechanism, participants could apply for small grants from the Institute to help them develop instructional materials and implement a training session in their home institutions. The next chapter presents more information on the grants as part of the discussion of post-Institute activities and implementation.

Chapter 5
Post-Institute Activities

This chapter provides a description of the activities undertaken by participants after the Institute in Jordan in September 2012 to implement what they had learned.

GRANTS TO INSTITUTE PARTICIPANTS

One of the requirements in the call for applications was that "Applicants *must* use one or more of the instructional materials developed at the Institute in their teaching in the [next] semester." Through a competitive Request for Applications (RFA), grants in the amount of $1,500 each were offered to enable participants to carry out these activities.[54] Of the 28 participants at the Institute, 23 applied for these grants either as individuals or as teams. Project staff reviewed the applications and awarded a total of eight grants based on the quality of application. The successful applications addressed the following issues: overall learning goals of the proposed activity; teaching methods to be used; expected audience; budget (including any in-kind support from the home institutions) and timeline; anticipated difficulties and how they might be addressed; and any attempts to sustain the teaching and promotion of responsible science in their home institutions and their country of residence.

Each of the successful grantees submitted a report about the resulting project and all but one attended the reunion to present and discuss the work. Four participants who did not receive funding also provided information about their implementation activities; one of these nonfunded projects also was presented at the reunion.

The various proposals (details in Table 5-1) related, to a greater or lesser extent, to the topics discussed at the Institute. The funded proposals called for introducing active learning techniques modeled at the Institute with topics related to responsible conduct of science; workshops on safe laboratory practices/biosafety; mentor-trainee relationships and responsibilities; misconduct/improper behavior; authorship; and ethical values in science and research. Some grantees collected data on their target audiences' knowledge of responsible science before and after implementing teaching modules on responsible science. Others focused on teaching how to design an experiment to implement a research project as a segue to discussing responsible conduct of science. The awardees focused on strengthening problem-solving abilities, enhancing critical thinking, and building capacity among educators. Some emphasized overall awareness of scientists' professional responsibilities. Within the context of the principal investigator's (PI's) country and institution, all proposals were to receive some form of institutional support (financial or in-kind) and all provided a plan for sustaining their projects.

[54] The application form may be found in Appendix G.

The materials created, sessions and workshops conducted, and results of these activities were the focus of the reunion meeting.

REUNION

The reunion meeting took place on April 20, 2013, at a hotel in Amman, Jordan, and on the campus of the Jordan University for Science and Technology (JUST) in Ibrid on April 21. Dr. Elizabeth Heitman represented the National Academies committee that planned the Institute along with project staff members Drs. Lida Anestidou and Jay Labov. Nine Institute participants from Algeria, Egypt, Jordan, and Yemen, seven of whom received grants to catalyze their implementation work after the Institute, attended the reunion, gave presentations about their individual projects, and contributed to the general discussions that serve as the basis for the remainder of this chapter and for the discussion and conclusions in Chapter 6. The names of participants, their institutions and countries, and a brief overview of their projects are presented in Table 5-1.

The Best Laid Schemes[55]: Several conference calls with the National Academies team that organized the reunion resulted in an agenda that was sent to participants before the reunion. Participants were scheduled to work in small groups on day 1, to

- Provide brief overviews of their projects to the other members of the group.
- Discuss aspects of their projects that worked as originally envisioned.
- Describe to other group members any surprises, new insights, and unexpected opportunities that presented themselves during and after their implementation activities.
- Expand on the kinds of assessments they used to measure what their intended audiences had learned and the evaluations they developed to determine the efficacy of their projects.
- Consider how their projects might be improved and sustained over time.

The second day at JUST was to be devoted to each group summarizing its discussions from the previous day, followed by full group discussion of surprises, insights, opportunities, assessments, and sustainability issues.

The organizing group thought that formal presentations should be short, with greatest emphasis on the other points for discussion noted above. However, all participants came prepared to describe their projects more extensively and to a person asked for the agenda to be altered to accommodate more detailed discussion of each individual project. Accordingly, day 1 and the morning of day 2 were devoted to detailed presentation and discussion of each project. The afternoon of day 2 focused on the sustainability of individual initiatives and consideration of how the Institute participants might collaborate both with each other and with members of the NRC organizing committee and staff to bring a much needed emphasis on teaching responsible science ("scientific integrity," as proposed by one Institute participant) to the Middle East–North Africa (MENA) region.

The lessons that the committee has drawn from the reunion are presented in the next chapter. Combined with other efforts to evaluate the Institute, these provide the basis for the committee's conclusions about next steps in its work in the MENA region and beyond.

---

[55] Burns, Robert. 1785. Poem: *To A Mouse.*

TABLE 5-1 Projects funded and/or presented at the reunion.

GRANT IMPLEMENTERS	PROGRAM FORMATS AND AUDIENCES	TOPICS	INSTITUTIONAL SUPPORT AND SUSTAINABILITY
ALGERIA			
Introduction to the Responsible Conduct of Research			
Halima Benbouza, Biotechnology Research Center (CRBt) Noureddine Yassaa, University of Sciences and Technology Houari Boumediene Abdelkader Bouyakoub, University of Oran (Es-Sénia) Ben Amar Cheba, University of Oran (Es-Sénia)	1-day workshop: - Interactive presentations with Q&A, presenter, and 3 facilitators - Brainstorming - Case studies - Audience response cards - Postcourse questionnaire 26 researchers/PhD students at the CRBt	Potential threats from biotechnology and life sciences: What is dual use research? Collaborative science Mentor and trainee relationships Being a responsible author Research integrity and misconduct	The CRBt offered facilities to hold the workshop. Implementers to propose that RCS course be included in the student curriculum of their institutions. Team also willing to give the RCS module in different scientific institutions.
EGYPT			
Future Perspective of "Responsible Conduct of Science" at South Valley University			
Mahmmoud Sayed Abd El-sadek, South Valley University Farag Khoday Moalla Hamed, South Valley University Hamdy Saad Sadek El-Sheshtawy, South Valley University	Five 3-hour workshops: - Interactive discussion - Case studies by the presenters and the audience - Brainstorming 45 faculty and graduate students on 3 campuses of South Valley University	Authorship responsibilities Scientific misconduct Responsible research practices	South Valley University offered infrastructure including seminar rooms and multimedia suppliers at the university's three campuses (Qena, Hurghada, and Luxor). The team plans to integrate

			this module into the current program "Improving Skills of Staff Members."
Principles of Professionalism in Science			
<u>Yahya Zakaria Eid</u>, Kafrelsheikh University	Two 3-hour workshops: - Lecture - Interactive discussions 35 academic staff and graduate students	Components of responsible conduct of research and science: - Authorship and publication practices - Plagiarism - Conflict of interest - Scientific integrity	Kafrelsheik University offered the conference hall and projector. New and more specialized teachable units will be developed to cover the needs of the participants. Negotiations with the university to allow these workshops to occur on a regular basis.
Teaching Safe Laboratory Practice in Mansoura University by Active Learning			
<u>Mohamed Mostafa Elhadidy</u>, Mansoura University Mohamed Salah El-Tholoth, Mansoura University	Learner-centered teaching course consisting of 4-5 blocks: - Brainstorming - Small group discussion: Safety scenarios, multiple choice questions - Poster design - Role play - Case studies 20 undergraduate students	Components of safe laboratory practice: - General safety practices - Laboratory hazards - Biological safety levels - Personal protective equipment - Spill response and waste disposal - Decontamination - Emergency response	The university provided seminar halls, facilities, and materials for the course. It also supported the time spent by the instructors to teach the course. The team plans to disseminate the teaching course to different research units as it grows and evolves

	2 90-minute facilitator training modules (4 facilitators)		based on the university's needs. They also plan to provide workshops on a regular basis for current and future faculty and students. They further plan to apply for grants to support their work.
Study of Responsible Conduct of Science in the Curricula of Scientific Schools			
Mohamed El-Sayed El-Shinawi, Ain Shams University 65 members of the Ain Shams Medical Students Research Association	2-day awareness campaign on the campus of Ain Shams University Medical School: - Booklet about RCR - Interactive maze with stations - Pre- and postcampaign questionnaire Lecture day 340 medical students (campaign) 185 medical students (lecture day)	Responsible conduct of research: - Animal welfare - Research misconduct - Protection of human subjects - Mentor-trainee relationships regarding authorship - Conflict of interest Lectures: - Medical research ethics - Cancer biology research lab: Progress and achievements - From operation room to research lab: Solving health problems and improving the quality of life	Ain Shams University provided lecture halls and space for the booths free of charge. The data collected from the questionnaires will be analyzed to inform the content of future workshops on RCS as well as an e-learning platform for students.
Responsible Conduct of Scientific Research			
Mohamed Labib Salem, Tanta University Marwa Ahmed Ali, Tanta University	Three 2-hour workshops/week, each devoted to one topic. The workshops will be repeated three times (2nd time in May and 3rd in June 2013)	Research integrity and misconduct Biosafety Mentor-trainee relationships and responsibilities	The proposed workshops will be integrated into the Faculty Leadership and Development Program (FLDP) that aims to

Yahya Ahmed S. Al-Naggar, Tanta University Amal Hashish, Tanta University Atef Nwair, Tanta University Eman Balah, Tanta University Soha Helmy, Tanta University	- Presentation - Open discussion - Small group work - Case studies 75 graduate students		enhance the skills of faculty members at all levels. The materials used during the course will serve as the foundation for core undergraduate and postgraduate courses in research responsibility, biosafety, and the mentor-trainee relationship in the university.
Interactive Learning for Teaching Nursing Administration Course*			
Yaldez K. Zein ElDin, Damanhour University	A block of four lectures in each of two semesters - Jigsaw - Small group discussion - Brainstorming - Role play 90-120 students/semester	Staffing Documentation Quality of patient care Scheduling	Use the outcomes from the evaluation to modify the lecture format, presentation and content Provide learning workshops for faculty
YEMEN			
Mentor-Trainee Relationships and Responsibilities			
Samira Al-Eryani, Sana'a University Huda Omer Basaleem, University of Aden Khaled Abdulla Al-Sakkaf, University of Aden Ahmed Moharem, Thamar	Two-day workshop - Presentations with active learning - Colored audience response cards - Small group work/discussion - Sharing discussions with other groups	Mentoring: origin of mentoring; about mentor-trainee qualities and relationships; what is mentoring; ethics of mentoring; Authorship Mentor trainee conflicts	The University covered 20% of the projected cost. The Center for Medical Education, Sana'a University will provide modest future

University Qais Abdullah Nogaim, Ibb University Fayza Hamood Eyssa, Sana'a University Abdusalem Mohammed Al-Mekhlafi, Sana'a University	- Case studies with role playing - Videos presenting cases 24 members of the academic teaching staff from the Faculty of Medicine and Health Sciences and training staff from the National Centre of Public Health Laboratories	Current situation of postgraduate regulations in universities in Yemen and mentoring programs	funding and venue to support future workshops.
Planning and Implementing Scientific Research			
Huda Omer Basaleem, University of Aden Khaled Abdulla Al-Sakkaf, University of Aden	4-day workshop - Backward design - Interactive learning - Case studies - Group discussion - Colored audience response cards 23 assistant and 2 associate professors from the Faculty of Medicine and Health Sciences	Planning and implementing scientific research: - Steps of the research process - Design of a scientific proposal - Research methods - Manuscript writing and publication Analysis of scientific misconduct Communicating research findings	The College of Medicine provided the venue, paid for the cost of 72 Internet hours, and covered the expenses for 3 facilitators. The team will continue to offer this training and incorporate the teaching methods in other courses.
*This activity was not funded by a project grant.			

Chapter 6
Evaluation, Insights, and Realities

This chapter presents a preliminary evaluation of the Institute.[56] It begins with an account of the final facilitator debriefing at the end of the Institute and also includes data from the survey (see Appendix G) sent to participants three weeks later and their open-ended comments about particular aspects of the meeting. Insights from the outcomes of the implementation grants and the discussions at the reunion are included as well. Finally, the committee offers its judgments, based on the experience of designing and implementing the Institute, to inform similar current and future activities.

EVALUATION

Final Facilitator Team[57] Debriefing

The facilitator team members met immediately after the Institute to share their thoughts about the event. The majority of the participants were committed to implementing the educational methods with a focus on responsible science at their home institutions. However, it was clear that many of the ideas introduced at the Institute were new to the participants. It also was clear that, in contrast to the National Academies Summer Institutes on Undergraduate Biology Education (NASI), a smaller amount of instruction about pedagogy per se (versus modeling pedagogy during discussions about responsible science) would be easier for participants to absorb and process.

Some of the Institute's potential impact was lost because of the lack of advance preparation. Unfortunately, the committee's expectation that participants would read the background materials prior to the Institute was not made clear. For future institutes it will be important to convey as clearly as possible everything what the participants are expected to do in advance. This also might include offering a series of questions or dilemmas to be considered during the Institute. Such questions, conveyed in cover letters or emails, rather than the background readings themselves, would engender greater interest and curiosity and alert participants to the kinds of problem solving to be undertaken during the Institute.

To be accepted to the NASI, one member per team was expected to participate in the reunion meeting during the academic year 2012-2013. The facilitator team agreed that it would be very important to (1) continue to provide all participants access to the Institute's materials

[56] As discussed in Chapters 3 and 4, in the field of education research "assessment" and "evaluation" are different concepts. Assessment refers to "tools for measuring progress toward and achievement of the learning goal" (Handelsman et al., 2007:19), while evaluation refers to "the process of analyzing the results of assessment and determining whether the goals have been achieved" (Handelsman et al., 2007:20).

[57] The "facilitator team" includes members of the committee and invited individuals who worked with participants at the Institute.

and (2) be available to help participants address questions after they return to their home institutions.

The facilitators and other leaders of NASI routinely identify participants who might be invited to serve as facilitators at future NASI. Identifying and preparing facilitators from the participant pool enables them to reflect on the goals, objectives, and implementation strategies from two perspectives and enables them to become leaders for disseminating NASI's goals and practices. The facilitator team concluded that a similar model would be appropriate for future Institutes.

Post-Institute Survey

As part of the evaluation process, the facilitator team developed a web-based survey for the participants. Three weeks after the Institute participants received an invitation to take the survey together with the Request for Applications for the implementation grants (see Chapter 5). Twenty-six of 28 participants responded to the survey, the results of which are described in the next several sections.

General Information about the Participants
General characteristics of the participants were discussed in Chapter 4 as part of the committee's approach to recruitment. Figure 6-1 shows that most participants identified themselves as university faculty while a few identified themselves as academic administrators.

The survey asked participants to indicate whether they teach primarily undergraduates, graduate or postdoctoral students, or others. The results are shown in Figure 6-2. Among the responses to the third choice ("Other") were demonstrator (i.e., a master's-level student), chief researcher, and faculty who teach both undergraduate and graduate students.

Participants were also asked to indicate up to three reasons they chose to attend the Institute. Figure 6-3 shows the percentage of each selected option.

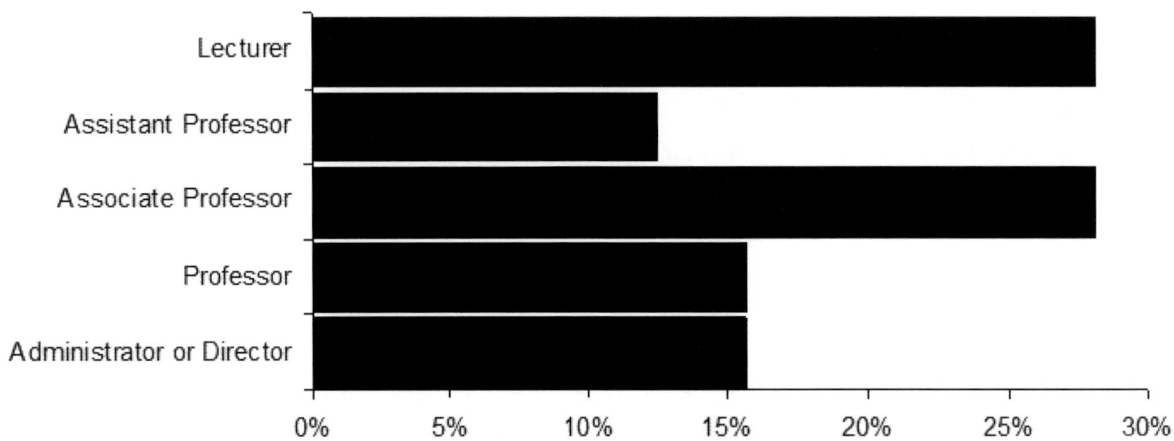

FIGURE 6-1 Rank of attendees by title (N=26). SOURCE: Data compiled by the committee.

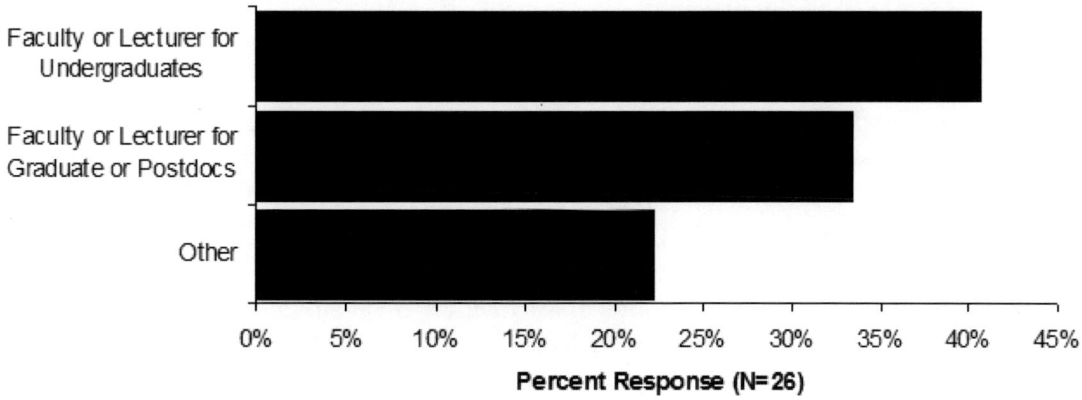

FIGURE 6-2 How participants described their primary audiences. "Faculty" refers to a participant who is actively engaged in research and who also teaches. "Lecturer" refers to a participant whose primary responsibilities are teaching. SOURCE: Data compiled by committee.

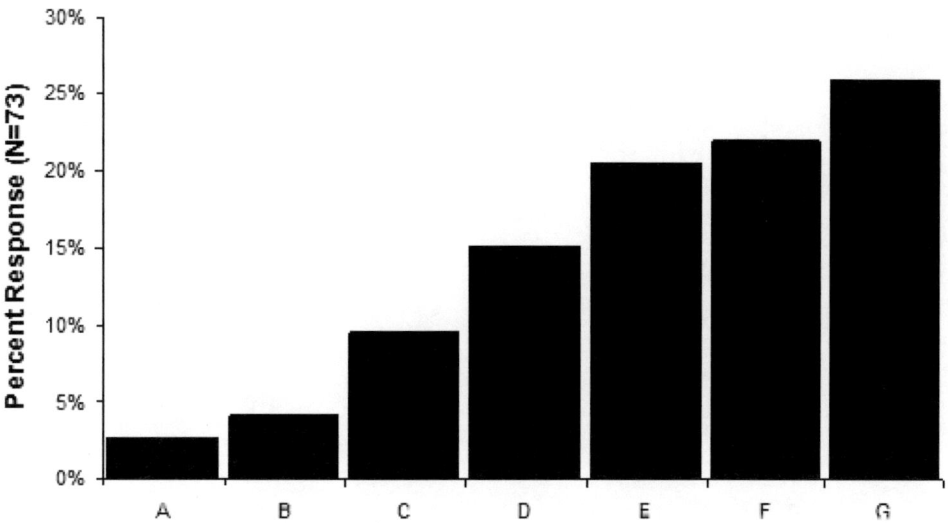

FIGURE 6-3 Reasons participants applied for the Institute. The answers in this figure are presented in ascending order starting with the option chosen by the fewest number of people. This differs from the order in which the options were presented in the survey.

A To reconnect with colleagues who share my interest in responsible conduct of science
B To meet colleagues from my country who share interests in responsible conduct of science
C To meet colleagues from other countries who share interests in responsible conduct of science
D To become more involved with future efforts to improve education about the responsible conduct of research internationally
E To deepen my understanding of the issues related to the responsible conduct of science
F To become more involved with future efforts to improve education about the responsible conduct of research in my country
G To discover tools, resources, and best practices for incorporating evidence-based teaching techniques into my courses

SOURCE: Data compiled by the committee.

Participants' Overall Rating of the Institute

Participants were asked to rate different aspects of the Institute; Figure 6-4 shows that more than 80 percent rated the quality of the sessions as either excellent or very good. There was a greater diversity of responses on questions about the use and balance of time spent in plenary and breakout sessions.

Figure 6-5 illustrates participants' high levels of satisfaction with the overall goals of the Institute, the instructional materials, and the relevance of the topics to their professional careers.

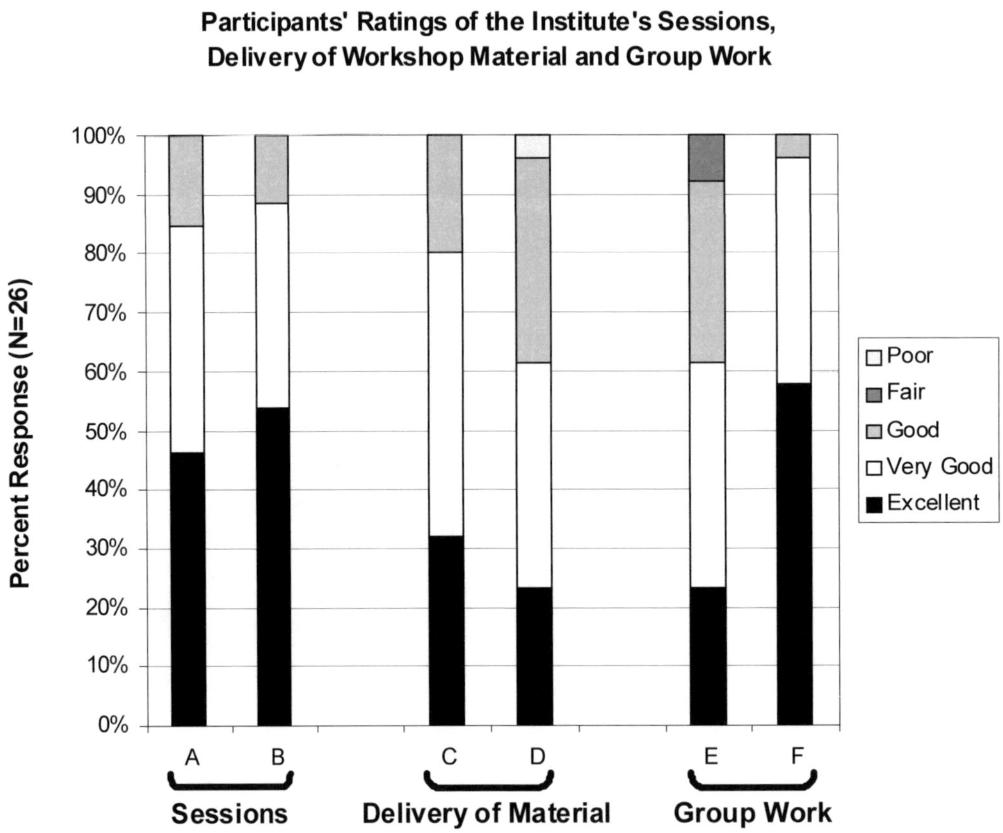

FIGURE 6-4 Participants' ratings of different aspects of the Institute, as defined:

- A Quality of sessions about the responsible conduct of science
- B Quality of sessions about the scientific basis for the use of active learning techniques
- C Inclusion of information and perspectives from a diverse range of views
- D Amount of time devoted to discussions during plenary sessions
- E Balance of time spent in whole group and team breakout sessions
- F Helpfulness of your breakout group's facilitators

SOURCE: Data compiled by the committee.

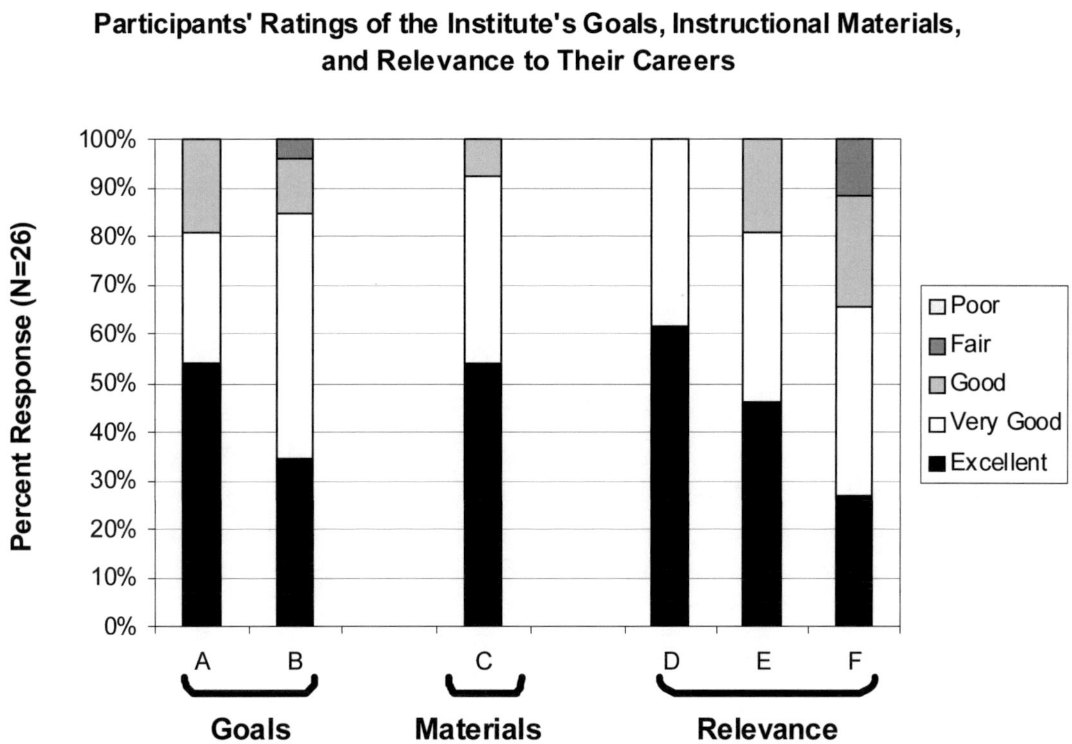

FIGURE 6-5 Participants' ratings of different aspects of the Institute, as defined below:

A Clarity of Institute's goals and objectives
B Relevance of topics that were presented in relation to the stated goals of the Institute
C Usefulness of resources provided by the organizers and presenters (e.g., background resources in the Dropbox and briefing book)
D Value of the Institute as a learning or professional development experience
E Relevance to you and your work of the issues presented
F Time to meet and interact with other participants
SOURCE: Data compiled by the committee.

When asked "If the National Academies were to organize and host additional Institutes or related activities on this topic in the future, would you be interested in participating?" 81 percent of the participants selected "definitely," with the remaining selecting "maybe." Of those who indicated they would like to be involved in future Institutes, 62 percent wrote that they would like to be a facilitator.

Open-Ended Comments

Participants were asked what they found to be particularly effective or not effective about the Institute. The majority of comments indicated that the Institute was effective for many reasons, but some reflected that the pace of the Institute was intense and the schedule crowded with too many subjects. Table 6-1 lists the participants' responses (edited for clarity), organized by effective and ineffective aspects of the Institute.

TABLE 6-1 Effective and Ineffective Aspects of the Institute. SOURCE: Information compiled by the committee.

Effective Aspects of the Institute

- Interactive comprehensive coverage of all topics in a friendly yet responsible environment.
- Round table discussion and cases are effective
- The open discussion was very effective...the organization of the groups at the beginning and during the workshop was great. Talks were unexpectedly awesome.
- Effective points: 1- Active interaction of well qualified trainers. 2- Time management. 3- Clear follow up plan. 4- Appropriate class facilities. 5- Hospitality.
- Everything was very interesting and very exciting: 1. Active Learning Techniques 2. Trainers 3. Scientific Material 4. Work in Teams 5. Exchange of Experiences
- Highly experienced faculty with simple transfer of data to participants
- Effective: Knowing other faculties nationally and internationally. The spirit of cooperation made the institute pass like one day.
- Smooth cruising into the presentation and discussion of the contents of the Institute and also dealing firmly and friendly from the institute presenters and facilitators with the participants.
- In my opinion all training sessions were effective.
- Conducting research responsibly; the development of professionalism in science; being part of the responsible scientific community.
- The committee and facilitators were serious and friendly at the same time. The use of all materials used in a manner not boring.
- The use of new approach in teaching and the use of dual science
- Most effective was Pedagogy
- Most of the activities in the institute were particularly effective.
- I found the effective points were the group discussion and how the facilitator helped us to get correct aspects and encouraged every participant to integrate with each other. Using the clickers during the lecture was new to me. How to teach the complex and difficult scientific topics in thoughtful ways.
- This is the first time I've attended such an intensive educational workshop. The tools such as case studies and role playing, I found more effective for me. The iclicker was also an effective tool to use for evaluation, however, I don't think I will use it at my institution with the large numbers of students....probably very expensive to get.
- I think the workshop was very valuable and gave me more experience and also gave me the chance to meet and deal with other international colleagues.
- The Institute was effective for many reasons: - It was an excellent training for me to be confronted to work with people from developed countries and countries who are facing the same problems as in my country. - To learn new tenets and pedagogical techniques for active learning. - Learn more about the different facets of what it means to conduct responsible science. - To share thoughts and learn on case studies about relevant topics: co-authorships, biosafety and biosecurity, international collaboration etc. - Develop a new network for future collaboration with mutual

interest/benefit

- New teaching techniques and the assessment methods I found particularly effective for me.
- I found that using case study and other methods of interactive learning was very effective and I will apply in teaching courses in my institution. Moreover, subjects of discussion like misconduct and safe laboratory standards are very important and direct our minds to very critical issues.
- They worked as one team and shared in all discussing points and activities in the workshop
- The responsible conduct of research thru discussing issues related to mentoring, authorship and active learning
- Interactive session on scientific misconduct cases
- J'ai sincèrement admiré le sérieux des organisateurs et facilitateurs et leur engagement dans le travail pour mener bien et réussir les objectifs qu'ils s'étaient fixés. Personnellement j'ai énormément appris sur le plan professionnel bien sûr mais aussi sur le plan humain ou j'ai vu à l'œuvre la générosité sans limite ni faille de certaines personnes, leur disponibilité à tout instant ainsi que leur penchant naturel à donner, à se rendre utiles sans pour autant espérer une contre partie. Tels furent à mes yeux les personnes qui ont pris en charge cette entreprise. Le groupe américain a était exemplaire à plus d'un titre... Que ses membres soient tous remerciés!
  (I sincerely admire the 'seriousness' (professionalism, effectiveness) of the organizers and facilitators and their commitment to successfully pursue the set goals. Personally, I learned an enormous amount at the professional level but also at the level of human relations observing in practice the limitless generosity and availability (of the aforementioned people). (I admired) their natural (spontaneous) offer to give (share) and to be of use without any compensation. Such have been the individuals in charge of this (whole) endeavor. The American group has been exemplary in more than one way (above and beyond the call of duty). All its members deserve (our) gratitude.)

Ineffective Aspects of the Institute

- The programme was very crowded.
- Dual use issues were delinquent.
- On the other hand, there were some issues regarding the place and the time of the workshop: (1) We took about 4 hours to travel from Amman to Aqaba and from Aqaba to Amman and this was fatiguing for me. (2) We start every day from (8 Am to 7 Pm) and this is too much time. (3) There is no entertainment and fun means during the workshop.
- The survey is ineffective.
- The intensive working hours is one drawback.
- I think the contents were very superficial as the institute tried to give us more than one subject in only one week as the pedagogy.
- Each topic should be a separate workshop.

INSIGHTS

NASI, which involves a variety of evidence-based approaches to active teaching, learning, engagement, and assessment, can be adapted to different topics, cultural contexts, and countries. In the course of reviewing the design and implementation of this Institute, the committee identified a number of insights that could help to improve future projects. They include logistical, academic, and cultural challenges and realities.

- Active engagement of committee members and Institute leaders before, during, and after the Institute is crucial.
- A detailed application and merit-based selection process can identify enthusiastic and committed participants who will, in turn, demonstrate the importance of such approaches to colleagues at their home institutions and in their disciplines.
- Teaching about and modeling pedagogy can play a significant role in the success of an Institute.
- The demanding pace of the Institute made it hard for some participants to comprehend the concepts and techniques fully and apply them during small group work. Future Institutes will benefit either by providing more time to integrate active learning with new content or by reducing the breadth or both.
- The design of resources and assessments for an Institute benefits from particular attention to linguistic and cultural differences among participants and facilitators. Working with partners from the region where the Institute will take place allows organizers to take into account local customs, traditions, and cultures in ways that remove barriers and foster stronger relationships among organizers and participants.
- NASI have demonstrated that a reunion of some participants after an Institute can provide new insights about participants' challenges, resources, and opportunities for networking and for sustaining programs (for details, see Chapter 5). The Institute described in this report further confirmed that a reunion can be especially important for participants from developing countries. For example, by the end of the reunion in Jordan, the scientists who attended agreed that their ability to conduct their own work around responsible conduct and to reach other colleagues at their home institutions, across their individual countries, and in the MENA region as a whole could be expanded and sustained by establishing a network among them. They decided to use this network to share ideas, common challenges, and opportunities, and to develop joint proposals for future work.
- As with the development of NASI, new Institutes will require continuing experimentation with and evaluation of all aspects of their design. Feedback from the participants, combined with the results of their projects, can play an important role in future iterations.
- The introduction of both new pedagogies and new content at the same time can be a significant challenge for some participants. Reviewing background materials in advance of the Institute can lessen this impact. However, materials written in English about new concepts, such as active learning and dual use, may present obstacles for non-English speakers.

REALITIES

- Framing biosafety and dual use issues in the context of responsible science was meaningful to many participants. However, based on conversations during plenary discussions with the participants who attended the reunion meeting in Amman, practical realities such as the lack of basic scientific equipment, reliable Internet connections, and access to scientific journals impede scientists in this region, and especially those from more impoverished nations, from undertaking research at a level where dual use issues raise concerns for them. People undertaking activities where research with dual use potential and/or misuse of technologies is to be a topic need to take this reality into account when planning their events or programs.

- Some concepts that are crucial to active learning, responsible science, and dual use cannot be expressed in Arabic. In most of the countries represented at this Institute, teaching about science occurs in English but instructors sometimes provide additional explanations or contexts in Arabic (or French in Algeria). Arabic-speaking scientists and students may interpret English words in ways that are different from what the organizers intend. For example, the facilitator team learned that there is only one Arabic word for the two English words "search" and "research," which may contribute to misunderstanding the standards for plagiarism in English-language journals among Arabic-speaking scientists and students. For example, several participants told the group that when they ask their students to define "research," the common response is to find the information in question on Google or another search engine. The students are not concerned about copying and pasting information from the Internet into their essays and research reports.
- Scientific research in the MENA region has advanced remarkably over the last generation. But participants reiterated that the lack of a formal framework and infrastructure for research in their countries (e.g., the absence of comprehensive policies and oversight structures regarding authorship, peer review, research with laboratory animals and human subjects, and biosafety) makes it difficult for scientists to follow international standards and to teach best practices in responsible science to their students.
- As the committee learned from the active learning exercise conducted on day 1 of the Institute, in which participants from each nation worked together to describe their country's system of higher education, there are both similarities and differences in education philosophies, approaches to teaching and learning, facilities, and resources among nations. These differences need to be taken into consideration when planning future Institutes. A recent report funded by the Carnegie Corporation of New York provides useful data to help organizers of future events take these differences into consideration (Bhandari and El-Amine, 2012).
- The small grants awarded to participants were used creatively to address an array of educational needs that they identified, as noted in Table 5-1. In many cases these funds prompted subsequent institutional support to sustain participants' instructional activities. However, as also occurs in the United States, limited funding restricted the ability of these motivated science educators to reach larger audiences who would benefit from instruction on responsible science, biosafety, and dual use issues.

At the reunion, discussions after each presentation and after all presenters had described their post-Institute activities revealed a great deal of variation in the ways participants in those activities were surveyed about their learning and the project's efficacy. Assessment and evaluation are an issue for science faculty around the world. Providing additional guidance and models of survey instruments before such projects are undertaken could provide much more useful and usable data for future initiatives.

Taken together, these insights offer important lessons for the design and implementation of future programs in the MENA region as well as in other parts of the world.

IMPLICATIONS: NEXT STEPS AND SUSTAINABILITY

As discussed briefly in Chapter 2, NASI, which are intended to transform how undergraduate biology is taught in the United States, have recognized that fundamental change takes time. Similarly, the committee agrees that for meaningful change to be sustainable, the projects and lessons learned from the first Institute need to be followed by additional efforts. In modifying future Institutes or similar activities, these efforts would also need to take into account the insights gained through the committee's evaluation work for the first Institute discussed above.

New Possibilities and Needs

Based on feedback from Institute participants, and others who became familiar with the Institute format as well as the committee, a series of ideas emerged about ways to reconfigure or extend the potential reach of the Institutes. The following four broad categories represent an amalgam of these suggestions:

- *Implications of dual use*
  The committee was charged with addressing research with "dual use" potential in the context of responsible conduct of science as part of its Statement of Task (Box 1-2 in Chapter 1). However, as a result of both designing the Institute and engaging with its participants, it became clear to the committee that the term "dual use" might not be the most appropriate one to use to communicate to the next generation of scientists and the various publics the complexity of the issues. Through the case studies presented and the discussions, it became apparent that "multiple uses" might be a preferable descriptor since virtually all scientific activities are on a continuum from exemplary to malicious conduct.[58] Given the differences between cultural norms, perspectives, and levels of scientific research among countries, scientists may be uncertain about boundaries of ethical/unethical behavior that "dual use" connotes because these behaviors are more complex than these two categories imply. There could, therefore, be value in emphasizing a continuum rather than a starker dichotomy of research and behavior as part of the discussions at the Institutes.

[58] The term was adopted by the International Union of Pure and Applied Chemistry for the educational materials on Multiple Uses of Chemicals that it developed in 2007 in cooperation with the Organization for the Prohibition of Chemical Weapons. The material, which was being updated when this report went to press, is available at *http://multiple.kcvs.ca/*. The IAC-IAP project on Responsible Conduct in the Global Research Enterprise chose the term "misuse" (IAC-IAP, 2012).

- *Involving policymakers and regulators in both the planning and conduct of future Institutes*
  This Institute focused on recruiting faculty and lecturers who would likely teach about the issues themselves. But policymakers and regulators of scientific policy, education, development, rules, and funding from the region where an Institute is conducted could contribute valuable insights, perspectives, and doses of reality for participants interested in developing RCS educational programs in their countries. In turn, policymakers and regulators could benefit from learning about the perspectives of scientists from their own and other countries in an environment that fosters respectful dialogue and challenges assumptions of individual participants.

  Consulting with policymakers and regulators from the region prior to an Institute also could help organizers to better understand and tailor the subjects and issues that they hope to address in ways that will be more meaningful to participants at an Institute. The importance of these connections became apparent when a committee member and a member of the project staff spent several days in Algeria consulting with representatives from various government offices as well as educators at Algerian universities in anticipation of a workshop there in June 2013 that is also sponsored by the Biosecurity Engagement Program at the U.S. Department of State. That workshop will assist Algerians in developing a national curriculum in bioethics.

- *Assessment of learning and evaluation of programs during institutes and in subsequent activities*
  As noted in Chapter 5 and earlier in this chapter, assessment of learning and evaluation of the efficacy of a program can be difficult because (1) people whose native language is not the one used to communicate may interpret words and phrases differently than the Institute organizers had intended, and (2) assessment of learning in higher education has traditionally been restricted to summative assessments that are given infrequently and are created by people with little expertise in psychometrics (the quantitative measurement tools and techniques developed in psychology). At the reunion meeting in Amman it became clear that grantees had used a broad spectrum of assessment and evaluation instruments. Participants were eager to know what instruments are already available that they could modify for their own purposes. Online instruments, such as the NSF-supported Student Assessment of Their Learning Gains (*www.salgsite.org/*), offer such templates. When developing future institutes, it would be helpful to provide a list of such resources and to spend some time helping participants understand their uses and value.

  For purposes of evaluating individual programs consistently, developing evaluation instruments that could be used by all participants who undertake subsequent activities could ease their workloads and make comparable data more readily available to Institute organizers.

- *Use of online technologies and resources*
  Institutes that involve regional or international travel for a small number of participants from any given country will, by themselves, have minuscule impact in addressing a very large set of national issues. NASI has begun to address this limitation by

expanding to a series of seven regional institutes each summer based on the annual institutes that were held in Madison, Wisconsin, for many years. However, regional institutes still cannot address the magnitude of change that is needed across hundreds of institutions and tens, perhaps hundreds of thousands of students. In addition, the costs for the Institute approach may be prohibitive in many parts of the world.

Thus the use of increasingly sophisticated online technologies and the development of online resources to reach much larger numbers of scientists, educators, and policymakers should be considered and supported. Social media, massive open online courses (MOOCs), and other forms of distance learning are some possible solutions. However, given this Institute's emphasis on evidence-based active teaching and learning, it must be recognized that overreliance on online technologies might compromise this aspect of the experience. A great deal of research is now under way to explore how such technologies might both enhance and compromise deep learning. The results of this work, in combination with the ability of web-based approaches to reach great numbers of students, should be taken into consideration by those who plan future programs.

Potential Next Steps in the MENA Region

A second regional Institute where the lessons from the first Institute would be applied is a logical activity to take advantage of the insights gained through this committee's evaluation work. As discussed in Chapter 2, one of the fundamental characteristics of successful faculty development programs is follow-up; no single event or experience is expected to be sufficient to foster genuine change. Such an Institute would not be a part of this National Academies project and would require new funding, although it could take advantage of the ties already created with institutions such as the Bibliotheca Alexandrina, The World Academy of Sciences (TWAS), or the Jordan University of Science and Technology (JUST).

The option for a follow-on Institute most favored by the committee would bring together several facilitators from the first Institute with some alumni. This arrangement could enhance their engagement with the methods and concepts promoted by the program. In the course of this second Institute, the Bibliotheca Alexandrina would have the opportunity to develop and host a website with materials from the Institutes as well as other resources to provide information and promote application of responsible science and active learning in universities and other research settings throughout the MENA region. Translating the Institute's relevant materials into Arabic would offer an outreach opportunity for interested scientists, policymakers, and others in the region.

It is also essential to help build participants' capacity to work more independently in their home countries. To begin the process, two participants from the first Institute, from Yemen and Egypt, will attend one of the 2013 regional Summer Institutes in the United States.[59] This weeklong immersion in active learning techniques will significantly increase their skills and abilities to implement active programs in responsible science in their own countries as well as to serve as facilitators at future Institutes. A number of participants already have envisioned one-half to two-day "mini-Institutes" to provide basic content and active learning

[59] *www.academiessummerinstitute.org/.*

experiences to a particular department or faculty. These might be logical projects for former Institute participants to create, perhaps in collaboration with a larger, continuing MENA participant network that involves the NationalAcademies in parallel with other comparable programs on responsible science and dual use issues.[60]

[60] For example, with regard to dual use issues, the programs on "dual use bioethics" operated by Bradford University and a new two-year, EU-supported project to create an "International Network of universities and institutes for raising awareness on dual-use concerns in bio-technology" that began work in January 2013 have connections to some of the countries in the MENA region. Further information about the Bradford activities is available at *www.brad.ac.uk/bioethics/about/* and about the EU project at *www.cbrn-coe.eu/Projects.aspx*.

References

2nd WCRI (World Conference on Research Integrity). 2010. Singapore Statement on Research Integrity. Available at *www.singaporestatement.org/*, accessed February 27, 2013.

AAAS (American Association for the Advancement of Science). 2008. Professional and Graduate-Level Programs on Dual Use Research and Biosecurity for Scientists Working in the Biological Sciences: Workshop Report. Washington. Available at *http://cstsp.aaas.org/files/AAAS_workshop_report_education_of_dual_use_life_science_research*.pdf; accessed March 11, 2013.

AAAS. 2011. Vision and Change in Undergraduate Biology Education. Washington.

African Union. 2006. The Cairo Declaration. Extraordinary Conference of the African Ministerial Council on Science and Technology (AMCOST), ET/AU/EXP/ST/Decl/13(II)\REV.1. 23-24 November. Available at *http://www.africa-union.org/root/au/auc/departments/hrst/biosafety/DOC/level1/EXT.AU.EXP.ST.DclI.13(II).Rev1.pdf*; accessed June 17, 2013.

Agre, P., and A. Leshner. 2010. Bridging science and society. Science 327(5968):921.

Albert, T. and E. Wager. 2003. How to handle authorship disputes: A guide for researchers. The COPE Report. Available at *http://publicationethics.org/files/2003pdf12.pdf*.

Beecher, H. 1966. Ethics and clinical research. New England Journal of Medicine 274(24):1354-1360.

Berg, P., D. Baltimore, S. Brenner, R.O. Roblin III, and M.F. Singer. 1975. Summary statement of the Asilomar Conference on Recombinant DNA Molecules. Proceedings of the National Academy of Sciences 72(6):1981-1984.

Bernhard, J. 2001. Does Active Engagement Curricula Give Long-Lived Conceptual Understanding? Pp. 749-752 in Physics Teacher Education Beyond 2000, R. Pinto and S. Surinach, eds. Paris: Elsevier.

Bibliotheca Alexandrina. 2012. Linking Science to Society: Proceedings of BioVision 2012. Alexandria, Egypt.

Bloom, B.S. 1956. Taxonomy of Educational Objectives: The Classification of Educational Goals. New York: D. McKay.

Branchaw, J., C. Pfund, and R. Rediske. 2010. Entering Research: Workshops for Students Beginning Research in Science. New York: W.H. Freeman & Co.

Bransford, J.D.E., A.L.E. Brown, and R.R.E. Cocking. 1999. How People Learn: Brain, Mind, Experience, and School. Washington:

National Academies Press.

Brewer, C., and D. Smith, eds. 2011. Vision and Change in Undergraduate Biology Education. Washington: American Association for the Advancement of Science.

CDC (Centers for Disease Control and Prevention). 2006. Reconstruction of the 1918 Influenza Pandemic Virus: Questions & Answers. Available at *www.cdc.gov/flu/about/qa/1918flupandemic.htm*; accessed March 11, 2013.

CDC and NIH (National Institutes of Health). 2007. Biosafety in Microbiological and Biomedical Laboratories, 5th ed. (L.C. Chosewood and D.E. Wilson, eds.). Washington: U.S. Government Printing Office.

CEN (European Committee for Standardization). 2008. International Laboratory Biorisk Management Standard. CWA 15793. Brussels.

Coley, J., and K.D. Tanner. 2012. Common origins of misconceptions: Cognitive principles and the development of biology thinking. CBE Life Sciences Education 11:209-215. Available at *www.lifescie.org/content/11/3/209.full.pdf/html*.

Cressey, D. 2007. Hippocratic oath for scientists. Nature News Blog. Available at *http://blogs.nature.com/news/2007/09/hippocratic_oath_for_scientist.html*; accessed June 4, 2013.

Crowe, A., C. Dirks, and M.P. Wenderoth. 2008. Biology in Bloom: Implementing Bloom's taxonomy to enhance student learning in biology. CBE Life Sciences Education 7(4):368–381.

Deer, B. 2011a. Secrets of the MMR scare: How the vaccine crisis was meant to make money. BMJ 342:c5258.

Deer, B. 2011b. Secrets of the MMR scare: The Lancet's two days to bury bad news. BMJ 342:c7001.

Deer, B. 2011c. How the case against the MMR vaccine was fixed. BMJ 342:c5347.

Enserink, M., and D. Malakoff. 2012. Will flu papers lead to new research oversight? Science 335:21-22.

Epstein, G.L. 2001. Controlling biological warfare threats: Resolving potential tensions among the research community, industry, and the national security community. Critical Reviews in Microbiology 27(4):321-354.

ESF (European Science Foundation). 2000. Good scientific practice in research and scholarship. Policy Briefing 10. Strasbourg, France. Available at *www.esf.org/fileadmin/Public_documents/Publications/ESPB10.pdf*; accessed June 4, 2013.

ESF. 2007. Research Integrity: Global Responsibility to Foster Common Standards. Policy Briefing 30. Strasbourg, France. Available at *www.esf.org/fileadmin/Public_documents/Publications/SPB%2030%20Research%20Integrity.pdf*; accessed June 4, 2013.

ESF. 2008. Stewards of Integrity: Institutional Approaches to Promote and Safeguard Good Research Practice in Europe. Strasbourg, France.

ESF. 2010. Fostering Research Integrity in Europe. Strasbourg, France.

ESF and ALLEA (All European Academies). 2010. The European Code of Conduct for

Research Integrity. Strasbourg, France.

EUBARnet (European Biosecurity Awareness Raising Network). 2012. Security Implications of Life Sciences and Technologies: A Multidisciplinary Approach. Prepublication copy. Como, Italy: Landau Network Centro Volta.

Fouchier, R.A.M., A. García-Sastre, Y. Kawaoka, W.S. Barclay, N.M. Bouvier, I.H. Brown, I. Capua, H. Chen, R.W. Compans, R.B. Couch, N.J. Cox, P.C. Doherty, R.O. Donis, H. Feldmann, Y. Guan, J.M. Katz, O.I. Kiselev, H.D. Klenk, G. Kobinger, J. Liu, X. Liu, A. Lowen, T.C. Mettenleiter, A.D.M.E. Osterhaus, P. Palese, J.S. Malik Peiris, D.R. Perez, J.A. Richt, S. Schultz-Cherry, J. Steel, K. Subbarao, D.E. Swayne, T. Takimoto, M. Tashiro, J.K. Taubenberger, P.G. Thomas, R.A. Tripp, T.M. Tumpey, R.J. Webby, R.G. Webster. 2013. Transmission studies resume for avian flu. Science 339 (6119):520-521; doi: 10.1126/science.1235140.

Garfinkel, M.S., D. Endy, G.L. Epstein, and R.M. Friedman. 2007. Synthetic genomics: Options for governance. Industrial Biotechnology 3(4):333-365.

Ghosh, P. 2007. UK science head backs ethics code. BBC News, September 12. Available at *http://news.bbc.co.uk/go/pr/fr/-/2/hi/science/nature/6990868.stm.*

Handelsman, J., D. Ebert-May, R. Beichner, P. Bruns, A. Chang, R. DeHaan, J. Gentile, S. Lauffer, J. Stewart, S.M. Tilghman, and W.B. Wood. 2004. Scientific teaching. Science 304(5670):521–522.

Handelsman, J., S. Miller, and C. Pfund. 2007. Scientific Teaching. San Francisco: Freeman and Sons.

Heitman, E., and R.E. Bulger. 2005. Assessing the educational literature in the responsible conduct of research for core content. Accountability in Research 12:207-224.

HHS (Department of Health and Human Services, Office of the Secretary, Office of Public Health and Science, Office of Research Integrity). PHS Policy on Instruction in the Responsible Conduct of Research (RCR). 2000. Available at *www.or.org/pdf/PHS_Policy_on_RCR.pdf.*

Hilborn, R.C., ed. 2013. The Role of Scientific Societies in STEM Faculty Workshops. College Park, MD: American Association of Physics Teachers. Available at *www.aapt.org/Conferences/newfaculty/upload/STEM_REPORT-2.pdf.*

Horton, R. 2004. The lessons of MMR. Lancet 363:747-749.

IAC (InterAcademy Council) and IAP (The Global Network of Science Academies). 2012. Responsible Conduct in the Global Research Enterprise: A Policy Report. Amsterdam.

IAP. 2005. Statement on Biosecurity. Available at *www.interacademies.net/CMS/About/3143.aspx*; accessed February 27, 2013.

ICMJE (International Committee of Medical Journal Editors). 2010. Uniform Requirements for Manuscripts Submitted to Biomedical Journals: Writing and Editing for Biomedical Publication. pp.1-3. Available at *www.icmje.org/urm_full.pdf.*

ICSU (International Council for Science). 2008. Freedom, Responsibility and Universality of Science. Paris.

IOM (Institute of Medicine). 1989. The Responsible Conduct of Research in the

Health Sciences. Washington: National Academy Press.

IOM. 2001. Preserving Public Trust: Accreditation and Human Research Participant Protection Programs. Washington: National Academies Press.

IOM. 2010. Ethical Issues in Studying the Safety of Approved Drugs. Washington: National Academies Press.

IOM. 2011. Chimpanzees in Biomedical and Behavioral Research. Washington: National Academies Press.

Jones, N. 2007. A code of ethics for the life sciences. Science and Engineering Ethics 13:25-43.

Kochan, C.A., and J.M. Budd. 1992. The persistence of fraud in the literature: The Darsee case. Journal of the American Society for Information Science 47(7):488-493.

Knight, J.K., and W.B. Wood. 2005. Teaching more by lecturing less. Cell Biology Education 4:298-310.

Kriesberg, N. undated. Animals as models. Available at *http://ori.hhs.gov/education/products/ncstate/models.htm*; accessed April 29, 2013.

Labov, J.B. 2004. From the National Academies: The challenges and opportunities for improving undergraduate science education through introductory courses. Cell Biology Education 3:212-214.

Labov, J., and J. Young. 2013. National Academies Summer Institutes on Undergraduate Education in Biology. In: Hillborn, R. (ed). pp. 21-24. The Role of Scientific Societies in STEM Faculty Workshops. Washington: Council of Scientific Society Presidents.

Maher, B. 2012. The biosecurity oversight. Nature 485:431-434.

Michael, J. 2006. Where's the evidence that active learning works? Advances in Physiology Education 30:159-167.

Meltzer, D., Thornton, R. 2012. Resource Letter ALIP-1: Active-Learning Instruction in Physics. American Journal of Physics 80(6).

Miller, S., and C. Pfund. 2011. Facilitator Training Packet. Madison: University of Wisconsin. Unpublished.

Morris, J. 2012. Did lax laboratory safety practices kill this UCLA chemist? Mother Jones. Available at *www.motherjones.com/politics/2012/07/ucla-sheri-sangji-patrick-harran*; accessed April 30, 2013.

Mukunda, G., K.A. Oye, and S.C. Mohr. 2009. What rough beast: Synthetic biology, uncertainty, and the future of Biosecurity. Politics and the Life Sciences 28(2):2-26.

NAE (National Academy of Engineering). 2007. "Ethics in the Science Classroom: An Instructional Guide for Secondary School Science Teachers." Online Ethics Center for Engineering. Available at *www.onlineethics.org/Education/precollege/scienceclass.aspx.*

Nature. 2012. Comment: The fight over flu. 481:257-259.

NIH (National Institutes of Health). 1990. Requirement for programs in the responsible conduct of research in national research service award institutional training programs. NIH Guide Grants Contracts. 19:1.

NIH. 1992. Reminder and update: requirements for instruction in the responsible conduct of research in NRSA institutional research training grants. NIH Guide Grants Contracts.

NIH. 2001. Notice of suspension of PHS policy. NOD-OD-01-020. Available at *http://grants.nih.gov/grants/guilde/notice-files/NOT-OD-01-020.html.*

NIH. 2009. Update on the Requirement for Instruction in the Responsible Conduct of Research. Notice Number: NOT-OD-10-019. Available at *http://grants.nih.gov/grants/guide/notice-files/NOT-OD-10-019.html*; accessed March 11, 2013.

NIH. 2012. Notice Pertinent to the March 2012 Revisions of the NIH Guidelines for Research Involving Recombinant or Synthetic Nucleic Acid Molecules (NIH Guidelines). Available at *http://oba.od.nih.gov/oba/rac/Guidelines/NIH_Guidelines_new.pdf*; accessed March 11, 2013.

NIH (Office of Extramural Research). 2013. Grants Policy: Research integrity. Available at *http://grants.nih.gov/grants/research_integrity/index.htm.*

Novossiolova, T., G. Mancini, and M.R. Dando. 2013. Toward Effective and Sustainable Biosecurity Education for the Life Sciences: The Benefits of Active Learning. Bradford, UK: Bradford Disarmament Research Centre.

NRC (National Research Council). 1989. On Being a Scientist. Washington: National Academy Press.

NRC. 1992. Responsible Science, Volume I: Ensuring the Integrity of the Research Process. Washington: National Academy Press.

NRC. 1995. On Being a Scientist, 2nd ed. Washington: National Academy Press.

NRC. 2000. How People Learn: Brain, Mind, Experience, and School (Expanded Edition). Washington: National Academy Press.

NRC. 2002. Scientific and Medical Aspects of Human Reproductive Cloning. Washington: National Academies Press.

NRC. 2003. BIO2010: Transforming Undergraduate Education for Future Research Biologists. Washington: National Academies Press.

NRC. 2004. Biotechnology Research in an Age of Terrorism. Washington: National Academies Press.

NRC. 2005. Guidelines for Embryonic Stem Cell Research. Washington: National Academies Press.

NRC. 2006a. Globalization, Biotechnology, and the Future of the Life Sciences. Washington: National Academies Press.

NRC. 2006b. America's Lab Report: Investigations in High School Science. Washington: National Academies Press.

NRC. 2007a. Science and Security in a Post 9/11 World: A Report Based on Regional Discussions Between the Science and Security Communities. Washington: National Academies Press.

NRC. 2007b. Taking Science to School. Washington: National Academies Press.

NRC. 2008. Ready, Set, Science! Washington: National Academies Press.

NRC. 2009a. A New Biology for the 21st Century. Washington: National Academies Press.

NRC. 2009b. Responsible Research with Biological Select Agents and Toxins. Washington: National Academies Press.

NRC. 2009c. On Being a Scientist, 3rd ed. Washington: National Academies Press.

NRC. 2009d. The 2nd International Forum on Biosecurity: Report of an International Meeting, Budapest, Hungary, March 30-April 2, 2008. Washington, D.C.: National Academies Press.

NRC. 2010. Research at the Intersection of the Physical and Life Sciences. Washington: National Academies Press.

NRC. 2011a. Guide for the Care and Use of Laboratory Animals, 8th ed. Washington: National Academies Press.

NRC. 2011b. Life Sciences and Related Fields: Trends Relevant to the Biological Weapons Convention. Washington: National Academies Press.

NRC. 2011c. Challenges and Opportunities for Education about Dual Use Issues in the Life Sciences. Washington: National Academies Press.

NRC. 2011d. Promising Practices in Undergraduate Science, Technology, Engineering, and Mathematics Education: Summary of Two Workshops. Washington: National Academies Press.

NRC. 2011e. Research in the Life Sciences with Dual Use Potential: An International Faculty Development Project on Education about the Responsible Conduct of Science. Washington: National Academies Press.

NRC. 2012a. Discipline-Based Education Research: Understanding and Improving Learning in Undergraduate Science and Engineering. Washington: National Academies Press.

NRC. 2012b. A Framework for K-12 Science Education: Practices, Crosscutting Concepts, and Core Ideas. Washington: National Academies Press.

NSABB (National Science Advisory Board for Biosecurity). 2007. Proposed Framework for the Oversight of Dual Use Life Sciences Research: Strategies for Minimizing the Potential Misuse of Research Information. Available at *http://oba.od.nih.gov/biosecurity/pdf/Framework%20for%20transmittal%200807_Sept07.pdf.*

NSABB. 2008. Strategic Plan for Outreach and Education on Dual Use Issues. Available at *http://oba.od.nih.gov/biosecurity/biosecurity_documents.html*; accessed March 11, 2013.

NSABB. 2010. Enhancing Responsible Science: Considerations for the Development and Dissemination of Codes of Conduct for Dual Use Research. Available at *http://oba.od.nih.gov/oba/biosecurity/documents/COMBINED_Codes_PDFs.pdf.*

NSF (National Science Foundation). 2009. Responsible Conduct of Research: NSF's Implementation of Section 7009 of the America COMPETES Act. Federal Register 74(160), 42126-42128. Available at *www.gpo.gov/fdsys/pkg/FR-2009-08-20/html/E9-19930.htm.*

The Nuremberg Code. 1949. Trials of War Criminals before the Nuremberg Military Tribunals under Control Council Law No. 10, Vol. 2. Washington: U.S. Government Printing Office, pp. 181-182. Available at *www.hhs.gov/ohrp/archive/nurcode.html*; accessed February 27, 2013.

OECD (Organization for Economic Cooperation and Development). 2004. Promoting responsible stewardship in the biosciences: Avoiding potential abuse of research and resources. Chairman's summary. Available at *www.oecd.org/sti/futures/long-termtechnologicalsocietalchallenges/33855561.pdf*; accessed March 11, 2013.

OECD. 2007. OECD Best Practice Guidelines on Biosecurity for BRCs (Biological Resource Centers). Paris.

OECD. 2009. The Bioeconomy to 2030: Designing a Policy Agenda. Paris. Available at *www.oecd.org/futures/long-termtechnologicalsocietalchallenges/thebioeconomyto2030designingapolicyagenda.htm*; accessed February 27, 2013.

ORI (Office of Research Integrity). 2000. Announcement of Final PHS Policy on Instruction in the Responsible Conduct of Research. Notice OD-01-007. Available at *http://grants.nih.gov/grants/guide/Notice-files/NOT-OD-01-007.html*; accessed June 4, 2013.

PCAST (President's Council of Advisors for Science and Technology). 2012. Engage to Excel: Producing One Million Additional College Graduates with Degrees in Science, Technology, Engineering, and Mathematics. Washington.

Pfund, C., S. Miller, K. Brenner, P. Bruns, A. Chang, D. Ebert-May, A. Fagen, J. Gentile, S. Gossens, I. Khan, J.B. Labov, C.M. Pribbenow, L. Tong, R. Wright, W.B. Wood, and J.A. Handelsman. 2009. Summer institute to improve biology education at research universities. Science 324:470-471.

Pfund, C., S. House, P. Asquith, K. Spencer, K. Silet, and C. Sorkness. 2012. Mentor Training for Clinical and Translational Researchers. New York: W.H. Freeman & Co.

Plionis, E. 2007. It's All the Little Things: How Misdemeanors in Scientific Misconduct Are as Bad as Fabrication and Falsification. Available at *http://serendip.brynmawr.edu/exchange/node/1636*.

Presidential Commission for the Study of Bioethical Issues. 2011. "Ethically Impossible": STD Research in Guatemala from 1946 to 1948. Washington.

Prince, M. 2004. Does active learning work? A review of the research. Journal of Engineering Education 93(3).

Rappert, B. 2003. Coding ethical behavior: The challenges of biological weapons. Science and Engineering Ethics 9(4):453-470. Available at *http://projects.exeter.ac.uk/codesofconduct/Publications/S&EE.htm*.

Rappert, B., ed. 2010. Education and Ethics in the Life Sciences: Strengthening the Prohibition of Biological Weapons. Canberra: The Australian National University E Press.

Rappert, B., and C. McLeish, eds. 2007. A Web of Prevention: Biological Weapons, Life Sciences and the Governance of Research. London: Earthscan.

Reverby, S.M. 2011. "Normal exposure" and inoculation syphilis: A PHS "Tuskegee" doctor in Guatemala. Journal of Policy History 23(1):6-28.

Revill, J. 2009. Biosecurity and Bioethics Education: A Case Study of the UK Context. Bradford, UK: Bradford Disarmament Research Centre.

Revill, J., and G. Mancini. 2008. Investigation into the Biosecurity Content of European Life Science Degree Courses. Como, Italy: Landau Network Centro Volta and Bradford Disarmament Research Centre.

Revill, J., G. Mancini, M. Minehata, and N. Shinomiya. 2009. Biosecurity Education: Surveys from Europe and Japan. Bradford, UK: Bradford Disarmament Research Centre.

Rickards, T., and Moger, S.T. 1999. Handbook for Creative Team Leaders. Brookfield, VT: Gower.

Rotblat, J. 1999. A Hippocratic Oath for scientists. Science 286:1475.

Royal Society and Wellcome Trust. 2004. Do No Harm: Reducing the Potential for the Misuse of Life Science Research. Available at *www.wellcome.ac.uk/stellent/groups/corporatesite/@policy_communications/documents/web_document/wtx023408.pdf.*

Royal Society. 2006. Report of the RS-IAP-ICSU International Workshop on Science and Technology Developments Relevant to the Biological and Toxin Weapons Convention. RS Policy Document no. 38(06).

Royal Society. 2012. Neuroscience, conflict, and security. Brain Waves Module 3. London.

Science. 2012. H5N1: Special section. 336:1522-1547.

Springer, L., M. Stanne, and S. Donovan. 1997. Measuring the Success of Small-Group Learning in College-Level SMET Teaching: A Meta-Analysis. Madison: National Institute for Science Education, University of Wisconsin. Available at *http://www.wcer.wisc.edu/archive/cl1/cl/resource/scismet.htm*; accessed June 17, 2013.

Steneck, N.H., and R.E. Bulger. 2007. The history, purpose, and future of instruction in the responsible conduct of research. Academic Medicine 82:829-824.

Stetson, N.E. 2009. Using small-group development to facilitate cooperative learning base groups. In J.L. Cooper, P. Robinson, and D. Ball, eds. Small Group Instruction in Higher Education: Lessons from the Past, Visions of the Future. Stillwater, OK: New Forums Press Inc.

Thornton, R.K. 2008. Effective learning environments for computer supported instruction in the physics classroom and laboratory. In: Connecting Research in Physics Education with Teacher Education, M. Vicentini and E. Sassi, eds. International Commission on Physics Education [online]. Available at *web.phys.ksu.edu/icpe/Publications/teach2/Thornton.pdf*; accessed June 17, 2013.

Tucker, J.B., and R. Zilinskas. 2006. The promise and perils of synthetic biology. The New Atlantis 12(1):25-45.

Tuckman, B. 1965. Developmental sequence in small groups. Psychological Bulletin 63(6):384-399.

UNESCO (United Nations Educational, Scientific and Cultural Organization). 1999. Declaration on Science and the Use of Scientific Knowledge. World Conference on Science, Budapest, Hungary, June 26-July 1. Available at www.unesco.org/science/wcs/eng/declaration_e.htm; accessed February 27, 2013.

UN (United Nations). 2004. A More Secure World: Our Shared Responsibility. Report of the High-level Panel on Threats, Challenges and Change. New York: United Nations.

van Aken, J. 2006. When risk outweighs benefit: Dual-use research needs a scientifically sound risk–benefit analysis and legally binding biosecurity measures. EMBO Reports 7(SI):S10–S13. doi: 10.1038/sj.embor.7400726.

Wakefield, A.J., S.H. Murch, A. Anthony, J. Linnell, D.M. Casson, M. Malik, M. Berelowitz, A.P. Dhillon, M.A. Thompson, P. Harvey, A. Valentine, S.E. Davies, J.A. Walker-Smith. 1998. Ileal-lymphoid-nodular hyperlasia, non-specific colitis, and pervasive developmental disorder in children [early report]. Lancet 351(9103):637-641.

WAME (World Association of Medical Editors). 2013. Publication Ethics Policies for Medical Journals. Available at *www.wame.org/resources/ethics-resources/publication-ethics-policies-for-medical-journals/*; accessed June 17, 2013.

Wheelis, M., L. Rózsa, and M. Dando. 2006. Deadly Cultures: Biological Weapons Since 1945. Cambridge, MA: Harvard University Press.

White House. 2012. National Bioeconomy Blueprint. Available at *www.whitehouse.gov/sites/default/files/microsites/ostp/national_bioeconomy_blueprint_april_2012.pdf*; accessed April 30, 2013.

WHO (World Health Organization). 2004. Laboratory Biosafety Manual, 3rd ed. Geneva. Available at *www.who.int/csr/resources/publications/biosafety/WHO_CDS_CSR_LYO_2004_11/en/*; accessed February 27, 2013.

WHO. 2005. Life Science Research: Opportunities and Risks for Public Health. Geneva. Available at *www.who.int/csr/resources/publications/deliberate/WHO_CDS_CSR_LYO_2005_20.pdf*.

WHO. 2006. Biorisk Management: Laboratory Biosecurity Guidance. Geneva. Available at *www.who.int/csr/resources/publications/biosafety/WHO_CDS_EPR_2006_6/en/*; accessed February 27, 2013.

WHO. 2007. Scientific Working Group on Life Science Research and Global Health Security: Report of the First Meeting. Geneva. Available at *www.who.int/csr/resources/publications/deliberate/WHO_CDS_EPR_2007_4*; accessed February 27, 2013.

WHO. 2010. WHO Biorisk Management Advanced Trainer Programme: Concept Sheet. Photocopy.

WHO. 2013. Report of the WHO Informal Consultation on Dual Use Research of Concern, Geneva, Switzerland, 26-28 February 2013. Geneva. Available at *www.who.int/csr/durc/durc_feb2013_full_mtg_report.pdf*, accessed May 14, 2013.

Wiggins, G., and J. McTighe. 1998. Understanding by Design. Alexandria, VA: Association for Supervision and Curriculum Development.

Wiggins, G., and McTighe, J. 2005. Understanding by Design, Expanded 2nd ed. Upper Saddle River, NJ: Pearson Publishing.

WMA (World Medical Association). 2008. WMA Declaration of Helsinki: Ethical Principles for Research Involving Human Subjects. Available at *www.wma.net/en/30publications/10policies/b3/*; accessed February 27, 2013.

World Science Forum. 2011. Declaration of the Budapest World Science Forum 2011 on a New Era of Global Science. Text adopted by the 5th Budapest World Science Forum on 19 November 2011 (definitive version). Available at *www.easac.eu/fileadmin/docs/WSF/WSF_Declaration_draft_19Nov_distribution.pdf*; accessed February 27, 2013.

Zilinskas, R.A., and J.B. Tucker. 2002. Workshop Report: Options for Limiting the Contribution of the Open Scientific Literature to the Biological Weapons Threat. Available at *www.hsdl.org/?view&did=718535*; accessed February 27, 2013.

Zimmerman, B.J., and D.H. Schunk. 2001. Self-regulated learning and academic achievement: Theoretical perspectives (2nd ed.). Mahwah, NJ: Lawrence Erlbaum Associates Publishers.

Zoller, U. 1993. Are lecture and learning compatible? Maybe for LOCS; Unlikely for HOCS. Journal of Chemical Education 70(3):195.

Glossary

Active learning instruction: combination of methods of instruction in which students are actively engaged in learning.

Alignment: ensuring that methods of active learning – including activities and assessments - will help students meet learning goals.

Assessments: methods and tools for gauging progress toward and achievement of the learning goals. Often categorized as formative and summative (see definitions below).

ALLEA: All European Academies, is a network of national academies of sciences and of humanities from Western, Central, and Eastern Europe. The current membership comprises 53 academiese from 40 countries; for more information see *http://www.interacademies.net/File.aspx?id=21281*.

Baby-boomers: people born between 1946 and 1964.

Backward design (reverse design): design of instructional materials and plans by first setting the learning goals, then determining what outcomes would reflect the attainment of those goals, and finally designing the aligned activities and assessments that will enable the accomplishment of the learning goals.

Bioethics: the discussion of controversial ethical practices brought about by advances in biology and medicine.

Biological containment: the combination of safety and security measures used to ensure that microorganisms capable of infection do not escape the research laboratory.

Biological Weapons Convention (BWC): disarmament treaty signed in 1972 that prohibits the development and stockpiling of biological and toxin weapons, along with the means of their weaponization and delivery. The BWC was the first international agreement to ban an entire class of weapons.

Biosafety: "the containment principles, technologies and practices that are implemented to prevent unintentional exposure to pathogens and toxins or their accidental release" (WHO, 2006:iii).

Biosecurity: "the protection, control and accountability for valuable biological materials[including information] … within laboratories in order to prevent their unauthorized access, loss, theft, misuse, diversion or intentional release" (WHO, 2006:iii).

Biosecurity Engagement Program (BEP): U.S. State Department Program "committed to developing cooperative international programs that promote the safe, secure and responsible use of biological materials that are at risk of accidental release or intentional misuse" (*http://www.bepstate.net/*).

Bloom's taxonomy: the six levels of cognition that represent a continuum of increasingly more conceptual learning tasks: knowledge, comprehension, application, analysis, synthesis and evaluation.

Code of Conduct: scientific organizations as well as governments have or support codes of conduct as one way to establish and promote responsible conduct, "thereby reducing threats associated with malign misuse of science, particularly areas associated with modern biotechnology" (Rappert, 2003).

Cognition: mechanisms that the brain uses to acquire and process knowledge and analyze information.

Cognitive science: the scientific discipline of the study of cognition.

Data fabrication: the presentation or publication of data that have not been generated through legitimate scientific processes or that are not supported by experimental results.

Data falsification: manipulation of data in any way that changes or omits data.

DBER: discipline-based education research, a collection of related research fields that study how students learn the knowledge, concepts, and practices of a particular discipline.

Dual use dilemma: The problem that arises in the life and other sciences because the same line of research could have the potential for great benefits but also for yielding knowledge, tools, or techniques that could be used to cause deliberate harm.

Dual use research: research intended for beneficial purposes that could nonetheless be misused for malevolent purposes.

Dual use research of concern: "Research that, based on current understanding, can be reasonably anticipated to provide knowledge, products, or technologies that could be directly misapplied to pose a threat to public health and safety, agricultural crops and other plants, animals, the environment, or materiel" (NSABB, 2007).

European Science Foundation (ESF): 72 member organizations dedicated to scientific research from 30 European countries comprise the European Science Foundation; for more information, see *http://www.esf.org/*.

Formative assessment: ongoing informal, low-stakes methods to provide information to both learners and instructors about next steps during the learning process.

Generation X: people born between early 1960s and the early 1990s.

Higher order cognitive skills (HOCS): complex judgment skills involving analysis, evaluation and creation of new knowledge (i.e., synthesis) as opposed to lower order cognitive skills (LOCS), or the learning of facts and concepts. LOCS typically correspond to the levels 1 -3 of Bloom's Taxonomy while HOCS correspond to levels 4-6 of Bloom's Taxonomy.

Human Genome Project: international scientific project with the primary goal of determining the entire DNA sequence (specific base pairs) and the estimated 20,000-25,000 genes encoded by those base pairs, on the 23 chromosomes of a human genome.

IAC: The InterAcademy Council, representing all of the world's science academies, "reports on scientific, technological, and health issues related to the great global challenges of our time, providing knowledge and advice to national governments and international organizations" (IAC in Brief [online]. Available at: *http://www.interacademycouncil.net/23450/27799.aspx*).

IAP—The Global Network of Sciences Academies: as one of its core activities IAP, which now includes over 105 national science academies, "works closely with its member academies to strengthen the role that science plays in society and to advise public officials on the scientific aspects of critical global issues" (About IAP [online] Available at: *http://www.interacademies.net/About.aspx*).

International Committee of Medical Journal's (ICJME) Uniform Requirements for Manuscripts: set of guidelines produced by the ICJME for standardizing the ethics, preparation and formatting of manuscripts submitted to biomedical journals.

Jigsaw exercise: cooperative learning strategy that enables each student of a "home" group to specialize in one aspect of a learning unit and then instruct and guide the other members of the home group. Each member of the group is essential to the completion of the unit.

Knowledge construction: learning theory developed by the education philosopher David Ausubel that proposes that learning builds upon and accommodates the experience of the learner, who integrates new knowledge into a personal framework or scaffold based upon those experiences.

Learning gains: "the percentage (or fraction) of the possible improvement that was actually achieved by students from pre to post-test, i.e., <g> = (Post - Pre)/(Perfect Score - Pre) x 100" (Thornton, 2008).

Learning goals: what students will know, understand and be able to do by the end of an instructional unit.

Lower order cognitive skills (LOCS): knowledge questions that require simple recall of information or simple application of known theory or concept; problems that can be solved without necessarily being understood.

MENA region: Middle East—North Africa region.

Meta-analysis: systematic method of integrating data from a number of studies addressing the same problem.

Metacognition: the process by which learners are aware of their levels of learning and, through that recognition, set learning goals, design approaches to achieve them, and monitor and evaluate progress towards the goals.

Millennial generation: people born between late 1970s and the early 2000s.

NIH: The National Institutes of Health, agency of the U.S. Department of Health and Human Services and the primary U.S. government agency responsible for biomedical and health-related research.

National Science Advisory Board for Biosecurity (NSABB): "a US. Federal government advisory committee chartered to provide advice, guidance, and leadership regarding biosecurity oversight of dual use research" (*http://oba.od.nih.gov/biosecurity/about_nsabb.html*).

NSF: The National Science Foundation, a U.S. government agency that supports fundamental research and education in all the non-medical fields of science and engineering.

OECD: Organization for Economic Cooperation and Development, an international organization dedicated to helping governments tackle the economic, social, and governance challenges of a globalised economy.

ORI: Office of Research Integrity, one of the bodies concerned with research integrity in the U.S. Department of Health and Human Services.

Plagiarism: appropriation of the ideas, language or expression of another. The precise delineation of an act of plagiarism is unclear and is considered culturally defined by some, although scientific standards with respect to publications and data do exist.

Recombinant DNA (rDNA): the transfer of DNA sequences from one organism to another by splicing or transplantation

Reverse-design: see backward design.

Risk: the potential that an activity or action may lead to a loss or some undesirable outcome.

Risk/benefit: the comparison of the risk of an action, activity or situation with its benefit.

Scaffolding: the framework of experience that learners use to organize and integrate new information in the process of knowledge construction and that instructors can provide to support learning.

Science of learning: research that seeks to understand learning at many levels of scientific inquiry, including physiology, neurology, psychiatry, psychology, cognition, sociology, developmental biology and genetics.

Scientific teaching: the pedagogical approach to the teaching of science that uses active learning methods and aligned assessments to measure learning with the same rigor as scientific research.

Soft law: "In the context of international law, soft law refers to guidelines, policy declarations or codes of conduct which set standards of conduct. However, they are not directly enforceable" (*http://definitions.uslegal.com/s/soft-law/*).

STEM: acronym for fields of study in the categories of science, technology, engineering, and mathematics.

Summative assessment: evaluation of student learning at the end of an instructional unit; such measures of accountability are generally used as part of the grading process.

Synthetic biology: "the design and re-design of biological parts, devices and systems" (*http://www3.imperial.ac.uk/systemsbiology*).

Synthetic DNA: artificially created strands of DNA made in the laboratory; the structure of the building blocks of DNA (4 bases with sugar and phosphates attached) are well understood and can be created de novo in the laboratory with increasing speed and lower cost.

Think, Pair, Share: activities that pose a question and allow students to consider the problem alone before discussing it with a classroom neighbor and then presenting conclusions to the class as a whole.

Transmissibility: the ability of an infectious agent to be passed from one host to another and cause disease.

Transmissionism: the tendency towards a more conventional, teacher-centered mode of instruction, with knowledge meant to be transmitted from teacher to pupil with little to no active learning methods involved.

UNESCO: United Nations Educational, Scientific and Cultural Organization, a body within the UN that encourages international peace and universal respect by promoting collaboration among nations.

WHO: World Health Organization, a specialized UN agency that is concerned with promoting international public health.

Appendix A
Recommendations from *Challenges and Opportunities for Education about Dual Use Issues in the Life Sciences*

This text is taken directly from NRC. 2011c. *Challenges and Opportunities for Education about Dual Use Issues in the Life Sciences*. Washington: National Academies Press, pp. 8-10.

SELECTED CONCLUSIONS

Educational Materials and Methods

The discussions during the workshop made clear that, beyond the available online resources, additional educational materials and resources are needed if discussions of research with dual use potential are to be incorporated more widely and effectively into education programs for life scientists around the world. Participants at the workshop addressed questions on the suggested content of these materials, the types of teaching methods that would be effective in presenting them, and the opportunities for developing materials more collaboratively and disseminating them more widely. One of the recurring themes in the discussion was that "no one size fits all," given the diversity of fields, interests, and experiences across the life sciences. The key is making the issue relevant to students and this requires a tailored approach. At the same time, participants also stressed the importance of finding ways to share successful practices and lessons learned as the scope and scale of education about dual use issues expands. The committee's conclusions with regard to these issues are:

- **Additional materials are needed that will be relevant to diverse audiences in many parts of the world, as well as those to at different educational stages, in different fields within the life sciences, and in related research communities. A number of good resources have been developed, but there is a need for more that are relevant to research related, for example, to plants or animals and to fields that are not as obviously security-related.**
- **More materials are needed in languages other than English. This will be particularly important in undergraduate settings or when used as part of technical training (i.e., biosafety).**
- **In addition to online resources, materials such as CDs or DVDs that can provide comparable opportunities for engaged learning are needed for areas that lack the sustained access or**

capacity to take full advantage of web-based materials.

- **Providing widespread access to materials that could be adapted for specific contexts or applications through open access repositories or resource centers would be important to implementing and sustaining education about dual use issues.**
- **Given current technology, it would be feasible to create the capacity to develop materials through online collaborations, as part of or in partnership with repositories or resource centers. Online collaborative tools can be a key mechanism to facilitate global participation in the development of materials, although again issues of access to the Internet will need to be considered in designing any arrangements.**
- **Developing methods and capacity for the life sciences and educational communities to comment on and vet education materials, such as an appropriately monitored Wikipedia model, would be important. Another important capacity would be the ability to share lessons learned and best practices about materials and teaching strategies as experience with education about dual use issues expands. If appropriate resources are available, both this and the previous conclusion should be well within the capacity of current online technologies.**
- **Teaching strategies need to focus on active learning and clear learning objectives, while allowing for local adaptation and application.**

Implementing Education about Dual Use Issues: Practical Considerations

A recurring theme during the workshop was the variety of settings in which content about dual use issues could be introduced. This reflected the diversity of the participants and the conditions in which education about dual use issues is currently taking place. It also led to discussions of a range of needs and challenges that are reflected in the committee's conclusions.

- **Incorporating education about dual use issues into the channels through which life scientists already receive their exposure to issues of responsible conduct—biosafety, bioethics and research ethics, and RCR—offers the greatest opportunity to reach the largest and most diverse range of students and professionals. Biosafety training reaches those with the most capabilities, knowledge, and motivation relevant to dual use. In addition, biosafety may be of particular interest for developing countries that are attempting to raise their overall standards of laboratory practices. Ethics and RCR are more general and may reach more people. The available evidence suggests that the use of multiple channels is already the most common approach.**
- **If the approach above is taken, then growing interest in expanding education about dual use issues, such as a proposal under consideration with the U.S. government to require such education for all federally funded life scientists, might also be an**

opportunity to expand more general education about responsible conduct.

- **It will be important to reach out to other disciplines that are increasingly part of life sciences research—physical sciences, mathematics, and engineering—as part of education about dual use issues. There may also be useful ideas and lessons from how these fields provide education about ethical issues and the potential for misuse of scientific results.**
- **Training opportunities to help faculty develop the skills, abilities and knowledge needed to teach dual use issues effectively are essential if education about dual use issues is to expand successfully.**
- **There are several promising models for "train-the-trainer" programs on which to draw, but a common characteristic is the use of the experience to create a network among faculty to support and sustain each other and to encourage expanded education.**
- **It is important to consider appropriate approaches to assessment and evaluation of education about dual use issues early in the process of developing and implementing new courses and modules.**
- **In addition to a lack of awareness of and engagement in dual use issues among life scientists, there are a number of obstacles to any effort to implement new content or teaching methods, such as competition for space in crowded curricula, pressures on students to focus on their research, and in some cases a general lack of support for teaching.**

RECOMMENDATIONS

General Approach

An introduction to dual use issues should be part of the education of every life scientist.

- **Except in specialized cases (particular research or policy interests), this education should be incorporated within broader coursework and training rather than via stand-alone courses. Appropriate channels include biosafety, bioethics and research ethics, and professional standards (i.e., RCR), as well as inclusion of examples of research with dual use potential in general life sciences courses.**
- **Insights from research on learning and effective teaching should inform development of materials, and approaches to teaching students and preparing faculty.**

Specific Actions

Achieving the broad goal of making dual use issues part of broader education will require a number of specific actions. They may be undertaken separately by different organizations but there will be substantial benefit if there is an effort to coordinate across the initiatives and share successful practices and lessons learned. Resources will be needed to ensure that the initiatives are carried out at an appropriate scale and scope.

The workshop participants and the committee did not explore the implementation of any specific recommendations in sufficient depth to prescribe a particular mechanism or path forward. Instead, reflecting the diversity and variety of situations in which education about dual use issues will be carried out, the final chapter lays out a number of options that could be used to implement each of the recommendations below.

- **Develop an international open access repository of materials that can be tailored to and adapted for the local context, perhaps as a network of national or regional repositories.**
  - **The repository should be under the auspices of the scientific community rather than governments, although support and resources from governments will be needed to implement the education locally.**
  - **Materials should be available in a range of languages.**
  - **Materials should interface with existing databases and repositories of educational materials dedicated to science education.**
  - **Additional case studies to address broader segments of the life sciences community should be developed, with a focus on making the case studies relevant to the student/researcher.**
- **Design methods for commenting and vetting of materials by the community (such as an appropriately monitored Wikipedia model) so they can be improved by faculty, instructors and experts in science education.**
- **Build networks of faculty and instructors through train-the-trainer programs, undertaking this effort if possible in cooperation with scientific unions and professional societies and associations.**
- **Develop a range of methods to assess outcomes and, where possible, impact. These should include qualitative approaches as well as quantitative measures, for example, of learning outcomes.**

Appendix B
The Bibliotheca Alexandrina
and
The World Academy of Sciences (TWAS), for the Advancement of the Developing World

BIBLIOTHECA ALEXANDRINA[61]

The New Library of Alexandria [inaugurated in 2002], the New Bibliotheca Alexandrina is dedicated to recapture the spirit of openness and scholarship of the original Bibliotheca Alexandrina. Its mission is

> to be a center of excellence for the production and dissemination of knowledge, and to be a place of dialogue and understanding between cultures and peoples.

It is our hope that the New Bibliotheca Alexandrina will be a worthy successor to the Ancient Library of Alexandria. That great Library was a unique ecumenical effort of the human intellect and imagination, and remains engraved in the memories of all scientists and intellectuals to this day.

The Ancient Library is undeniably the greatest chapter in the history of Alexandria. Our great city, founded by Alexander and home to Cleopatra, has had a remarkable history of 2300 years. It is a city of living history and renewed imagination that has inspired creative talents from Callimachus to Lawrence Durrell. In addition, the past is suddenly coming alive as underwater archaeology is bringing to light the sunken treasures of Alexandria, capturing the imagination of the world with glimpses of bygone glory.

That is the setting for the New Library of Alexandria. The beautiful new building, with its distinctive granite wall covered by the letters of all the world's alphabets, is today a recognizable landmark of the new Alexandria.

Before we turn to the future, it is only fitting that we should salute all those whose vision and dreams launched this great enterprise more than quarter-of-a-century ago, from UNESCO to the architects and engineers, and contractors, from the management of the project to the workers who labored in the quarries, from the Associations of Friends of the Library all over the world to the eminent people who served on international commissions, from the generous Government donations to the many individual donations. All must be thanked for having brought us to this important achievement.

The four objectives of the New Bibliotheca Alexandrina are to be:

1. The window of the world on Egypt;

[61] This material is taken from the "Director's Message" on the Library's website (*www.bibalex.org/aboutus/message _en.aspx*.

2. The window of Egypt on the world;
3. An instrument for rising to the digital challenge;
4. A center for dialogue between peoples and civilizations;

The way forward is difficult and challenging. The Library seeks to establish itself as an international center of excellence. In terms of our collections strategy, we focus on: First, the Ancient Library of Alexandria, Alexandria and Egypt; Second, the Mediterranean, the Arab world (without duplicating other efforts underway) and Africa, then the rest of the world. In terms of thematic focus, the Bibliotheca Alexandrina's specialized centers and departments undertake a number of specific projects and activities which complement and support one another in a coherent fashion. These projects all contribute to the BA's mission.

The means to move forward is partnering with many eminent institutions of learning around the world, either in an ongoing manner or around specific events such as seminars, conferences and exhibitions. Equally important to these links with eminent institutions are the links to the civil society in Egypt and the world. It is here that the 34 Associations of Friends of the Library have an invaluable role to play.

It is also challenging to link up electronically with the rest of world. We have already put together a complex web of agreements to bring the marvels of the digital age to all parts of Egypt and the region, and to bring the fruits of Egyptian creativity and scholarship to the new digital world of instant communications and electronic publishing.

Supported by the Council of Patrons, guided by the Board of Trustees, and in constant touch with the Friends of the Library of Alexandria, in Egypt and all over the world, the staff of the Bibliotheca Alexandrina are moving forward to build, over the years to come, an institution worthy of bearing that great name. We hope it will indeed be "a source of pride for Egypt and the world".

Ismail Serageldin
Librarian of Alexandria

THE WORLD ACADEMY OF SCIENCES (TWAS)[62]

TWAS is an autonomous international organization, founded in 1983 in Trieste, Italy, by a distinguished group of scientists from the South under the leadership of the late Nobel laureate Abdus Salam of Pakistan. It was officially launched by the secretary-general of the United Nations in 1985.

TWAS represents the best of science in developing countries. Its main mission is to promote scientific excellence and capacity in the South for science-based sustainable development.

The Academy's strength resides in the quality and diversity of its membership -- internationally renowned scientists elected by their peers. TWAS Fellows, who live and work in developing countries, represent 85 percent of the membership; TWAS Associate Fellows live and work in developed countries. The current membership stands at 1073 *[15 January 2013]*.

A Council, elected every three years by TWAS members, is responsible for the Academy's broad policy and programmatic directions. The Secretariat, headed by an executive director and located on the premises of the Abdus Salam International Centre for Theoretical Physics in Trieste, Italy, assists the

[62] The material is taken from the TWAS website (*http://twas.ictp.it/about/whats-twas*). The new name of the organization, formerly the Third World Academy of Sciences and then briefly TWAS, the academy of sciences for the developing world, was adopted in September 2012.

Council in the administration and coordination of the programmes.

In 1991, the United Nations Educational, Scientific and Cultural Organization (UNESCO) assumed responsibility for administering TWAS funds and personnel on the basis of an agreement signed by TWAS and UNESCO. In 2004, the Italian government passed a law that ensures a continuous financial contribution to the Academy's operation. Representatives of the Italian government and UNESCO are members of the TWAS Steering Committee, which meets annually to discuss financial matters.

In addition to its strong links with UNESCO and ICTP, TWAS provides administrative support for the Organization of Women in Science for the Developing World, IAP—The Global Network of Science Academies, and the InterAcademy Medical Panel). The Academy also maintains close ties with academies, research councils and ministries of science and technology in developing countries.

Objectives

- Recognize, support and promote excellence in scientific research in the developing world;
- Respond to the needs of young scientists in S&T-lagging developing countries;
- Promote South-South and South-North cooperation in science, technology and innovation;
- Encourage scientific research and sharing of experiences in solving major problems facing developing countries.

Appendix C
Detailed Results of the June 2011 Planning Meeting

This text is taken from the letter report of the meeting (NRC. 2011e. *Research in the Life Sciences with Dual Use Potential: An International Faculty Development Project on Education about the Responsible Conduct of Science.* Washington: National Academies Press, pp. 14-19). The material has been lightly edited to ensure that references to boxes or tables or specific pages are appropriate for this report.

GENERAL CHARACTERISTICS OF THE EGYPTIAN PROTOTYPE INSTITUTE (EPI)[63]

Advance planning. Since this is a new endeavor for the National Research Council (NRC), the preparations for the first workshop included the formal planning meeting and a site visit. If the program is successful, it is assumed that other countries in the MENA region will be able to participate in workshops hosted by the Egyptian network as the basis for launching their own projects. The NRC may have a supporting role but there will be less hands-on involvement as countries gain experience and take "ownership." This is the model that the National Academies Summer Institutes (NASI) program has adopted as it expands from a single national institute to multiple regional ones (see Chapter 3). There may still be cases where an initial site visit would be helpful, for example when the program begins in a new region, but the intent is to build a largely self-sustaining endeavor.

[63] This is the title adopted when it was assumed the focus would be on a single country. With the move to a regional approach, the title of the institute became *Education in responsible research with infectious diseases: Ensuring safe science in the 21st century.*

The Workshop Itself

The success of the NASI program (Pfund et al., 2009), as well as of other programs for faculty development, have suggested some basic features for a workshop:

- In person. Although it is becoming increasingly feasible to create and sustain virtual networks using resources such as videoconferencing and web 2.0 communications, there is still substantial value in bringing people together to be immersed in a common experience. Personal interactions also allow for informal communication outside the defined schedule that can be valuable to the network-building process.
- Duration. Experience from 8 years of NASIs suggests that 4 to 5 day long workshops would be optimal, given the amount of new material that participants would be expected to absorb and the value of cumulative learning-by-doing (see Chapter 2). 3Participants would be expected to do some advance preparation, but the main

experiences would be obtained during the meeting itself.

- Team-based. A key element for ensuring success and enhancing sustainability in the NASIs is the participation of teams from institutions, preferably including a range of junior to senior members on each team. Gaining buy-in from administrators is critical and it has proved useful to have them among the participants. The NASI model has shown added success and commitment by participants if their home institute provides at least modest resources to help implement what faculty learn.
- Hands-on. As the design of the planning meeting suggested, the workshop would be built around extensive, direct participation. Participants would have the opportunity to be both "students" and "teachers," to practice the methods they are learning, and to develop "teachable tidbits" and other materials (e.g., appropriate assessments) to help them implement their new courses or modules at their home institutions.
- Implementation and Assessment. An important feature of the workshop's hands-on approach is the commitment to assist participants in implementing what they have learned. In addition to implementing new ideas or courses, they will acquire experience and resources to plan and carry out effective assessments of whether the learning goals of their new activities are being met. As already mentioned in the context of sustainability, thinking about assessment from the outset is helpful on multiple levels. Examples of useful assessment techniques include observation of the participants, collecting and analyzing work samples, introducing checklists of skills, use of quizzes and/or self-assessment tools, interviews, etc.

The Network

Fostering successful and sustainable networks of faculty able to teach about dual use issues and broader problems of responsible conduct in science and research depends on several key elements, some of which have already been discussed earlier in this report.

- From the beginning. Given the emphasis on forward planning, strategies for building and sustaining the network of faculty will be part of the earliest discussions of the workshop. As previously presented, networks will be influenced by the local/national context, for example with regard to the degree of faculty autonomy in course design.
- Resources. As mentioned above, whenever possible participants in the workshop will be provided with materials and other resources to help them implement what they have learned. Modest resources from their home institution to show its commitment and obligation may be particularly desirable in the project's initial stages. It is the existence and ready availability of these resources rather than their amount that matters most; in many situations modest resources can have a significant impact.
- Continuing connections. Another way to help build a network is to have project staff from the sponsoring organization available for consultation to participants after the workshop as they implement their new ideas (courses, modules, etc.). These connections would reinforce rather than substitute for local commitment.
- Appraisal. The NASI arranges for at least some of the team members to get together approximately six months after the Institute to share experiences and challenges, reinforce ties, and make plans and adjustments. This is always important but is

particularly critical in the early days of a long-term project, i.e., the first years of implementation. The anticipation of a reunion may also encourage participants to persevere with applying their new skills, since it should be expected that, in spite of resources and support, at least some of them would encounter barriers or become discouraged.

DETAILS OF THE EGYPTIAN PROTOTYPE INSTITUTE GOALS AND LEARNING OBJECTIVES

The syllabus (e.g., content and pedagogy) of the institute is developed in close consultation with the faculty in whose country it will take place. The elements described below have been adapted to the needs identified by the faculty from research institutions in Egypt. Consequently, these may have to be modified to best fit the characteristics of each country.

During the planning meeting in Trieste, the general themes of the EPI were identified but the detailed content was not discussed. This is one of the tasks that the Committee overseeing this project is working on in close collaboration with the experts from Egypt who took part in the planning meeting.

The Importance of the Workshop's Title

In the planning meeting a substantial amount of time was devoted to selecting an appropriate title for the future Institute. While the chosen title reflects the core interests of the planners, it was mostly shaped by the Egyptian experts. It is aspirational and evokes the notions of education; responsible research; infectious diseases (or other life science); and safety in science: *Education in responsible research with infectious diseases ensuring safe science in the 21st century.* It also reflects the sensitivities to concepts such as dual use and biosecurity under current conditions in Egypt; it is unclear whether other workshops in other settings would experience the same concerns as strongly.[64]

Goals of the EPI

Expanding on the themes previously discussed, the following three are the goals to achieve by the faculty workshop:

1. *Understand the ethical and legal responsibilities of physical and life scientists.* The existence of multinational and multidisciplinary perspectives on what constitutes responsible life sciences research makes a discussion on the various norms and cultures of the practice of science very valuable. It would also foster the idea of a global science and research community, although the amount of legal information necessary is a matter of discussion. At the end of the workshop the participants will have a clearer appreciation of responsible conduct in research and science.
2. *Educate participants in the conduct of responsible science.* The workshop will foster good practice in teaching life and physical sciences and teach participants to adapt these to their own subject matters. At the end of the workshop the participants will have an appreciation for active learning techniques as these apply to responsible scientific practices, they will be able to utilize the teaching methods of the workshop, and to incorporate the workshop materials into existing programs in their own institutions.
3. *Cultivate future leaders in responsible science and research integrity.* In order to sustain the

[64] See NRC (2011c) and Rappert (2010) for accounts of the experiences of programs on dual use issues in other countries.

impetus for this project and foster a sense of achievement and dignity the workshop participants will be encouraged to not only develop good research practices but to identify the necessary support system to facilitate such changes. In the formative years of the project, the accomplishments of the site visit and the guidance of the NRC Committee members will be crucial to identify champions and to foster the exchange of scientists around the world to sustain this effort.

An example of how to structure the activities at the institute using a learning outcomes approach is shown in Table C-1.

Activities and Assessments

There are numerous activities to choose from to implement what was learned at the EPI at each participant's home institution. The choices could be influenced by what integrates well within a laboratory, a department or an institution and what is commonly used and accepted in a country's educational system. Pfund and colleagues have described a number of activities originating from the 6 years of Summer Institutes (Pfund et al. 2009), and below are some additional examples:

- Brown bag seminars
- A new course on responsible conduct of research (this may take a long time for approval, depending on the national structure of education curricula in a country)
- Incorporation of new teaching methods within existing courses in the life sciences adding the elements of RCR/RI teaching

At the end of the project a meeting of the EPI participants, Committee members and project staff will take place to measure success, discuss challenges and new activities to be undertaken (this also happens with the NASI). While no specific assessment tool has been designed, oral deliberations –especially during the formative years of the project- between participants are thought to be the most helpful assessment tool. It is possible that, following the completion of the EPI and the debriefing meeting a few months later, the Committee will formulate guidelines on data to be collected from participants and analyzed in the footsteps of the NASI.

Costs and Implementation Issues

Although these are important issues, they can only be addressed after the EPI has taken place.

TABLE C-1 Example of a "Learning Outcomes" approach.

General goals addressed	Specific learning objectives/outcomes	Types of assessments that measure objective	Activity that accomplishes that specific objective
Participants will be advocates for teaching responsible conduct of research and practice of science.	Develop a teaching module to illustrate the use of the concepts of responsible conduct of research.	Develop an assessment instrument that will demonstrate the student's ability to use the concepts you have discussed to solve practical problems. Use a historical case study to engage students and deepen their awareness of the various issues.	Present your approach to your colleagues in the Institute and obtain their feedback.
Participants will have an awareness of hazards in the laboratory and know how to bring that awareness to others.	Identify the difference between chemical and biological hazards. Be able to describe biosafety guidelines and standards of practice to prospective trainees	Tested knowledge; pre- and postassessment. Offer a problem and ask students to describe any obvious hazardous situations.	Group activities, small group discussions, clicker questions. Expertise sharing (own experiences of best practice; own stories of not-so-best practices).
Appreciate the ethical, legal, and social responsibilities of life scientists.	Indentify polices and guidelines and regulatory statements of both international and local bodies and critique the applicability of these statements. Able to write standards of practice for their own institution, department, or laboratory.	Convey these policies to the workers/students in their native language. Critique and discuss how these apply to participants' own experience, laboratory, institution, or country.	Locate and read/discuss these guidelines with the group. Discuss cases from historical examples (e.g., Thomas Butler). Discuss case studies specific to the group itself, e.g., based on personal experience.

Appendix D
Active Learning Toolkit and Images

This appendix contains materials intended to illustrate and supplement the discussions of active learning in the text of the report. It includes:

- Images, including cartoons, that illustrate active learning concepts and applications;
- Projects and resources devoted to promoting active learning in science education; and
- References to research on the science of learning, expanded from the references cited in the text.

USEFUL IMAGES

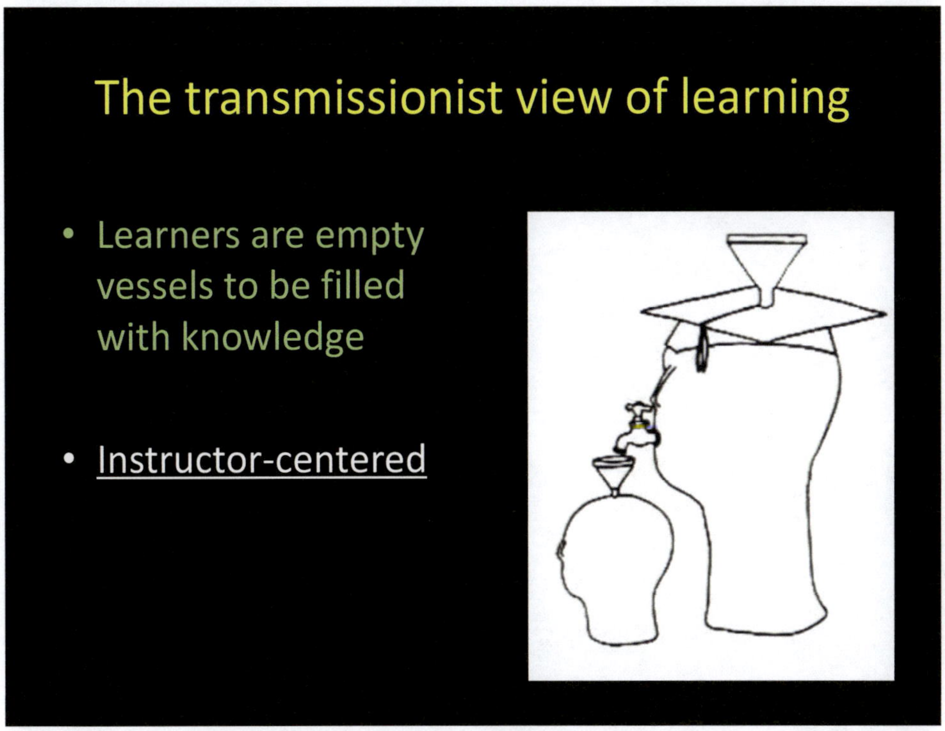

From Smith et al. 2005. *Journal of Engineering Education*. Used with permission.

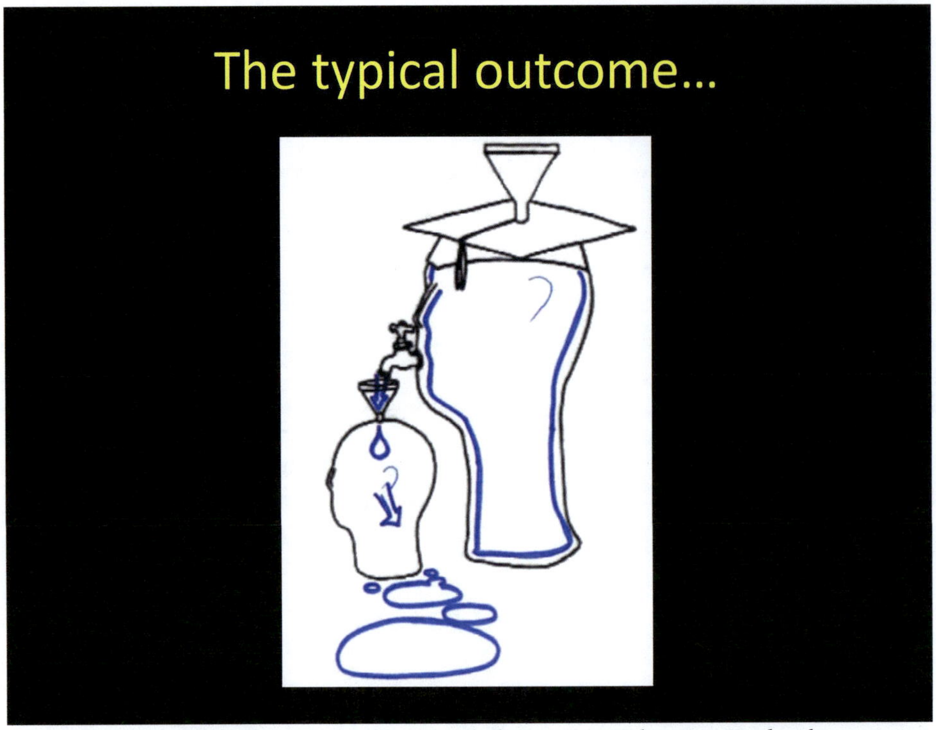

Adapted from Smith et al. 2005. *Journal of Engineering Education*. Used with permission.

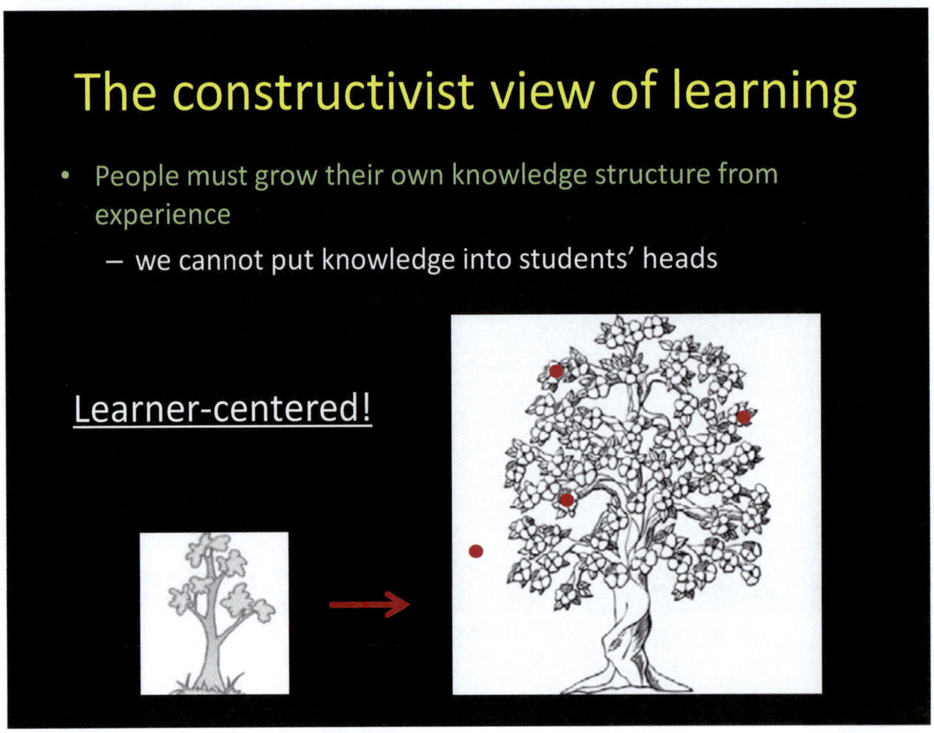

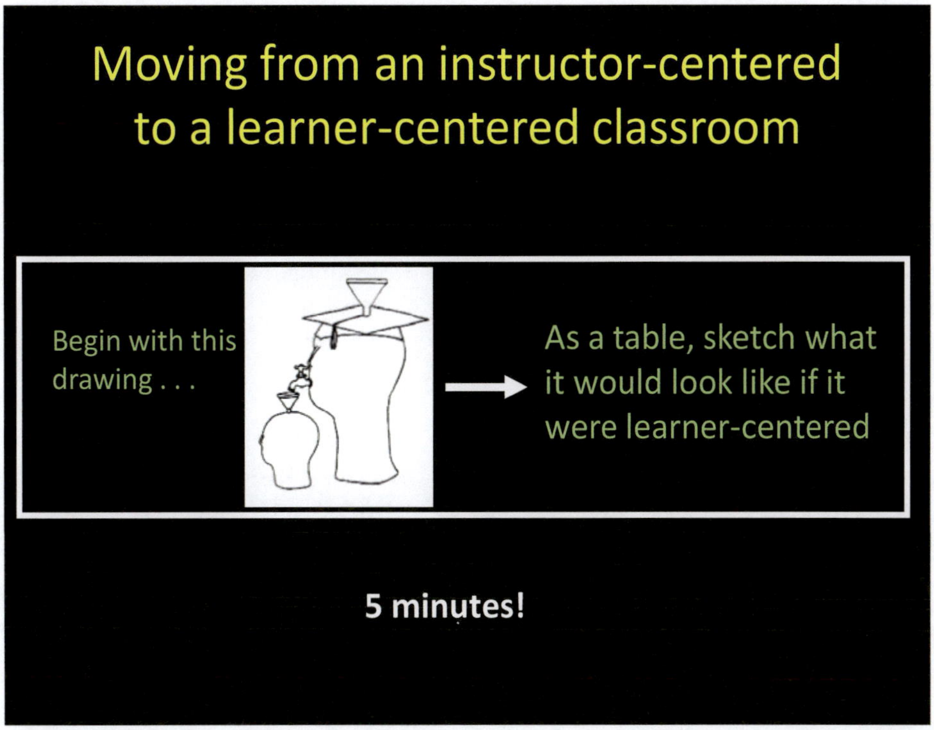

Adapted from Smith et al. 2005. *Journal of Engineering Education*. Used with permission.

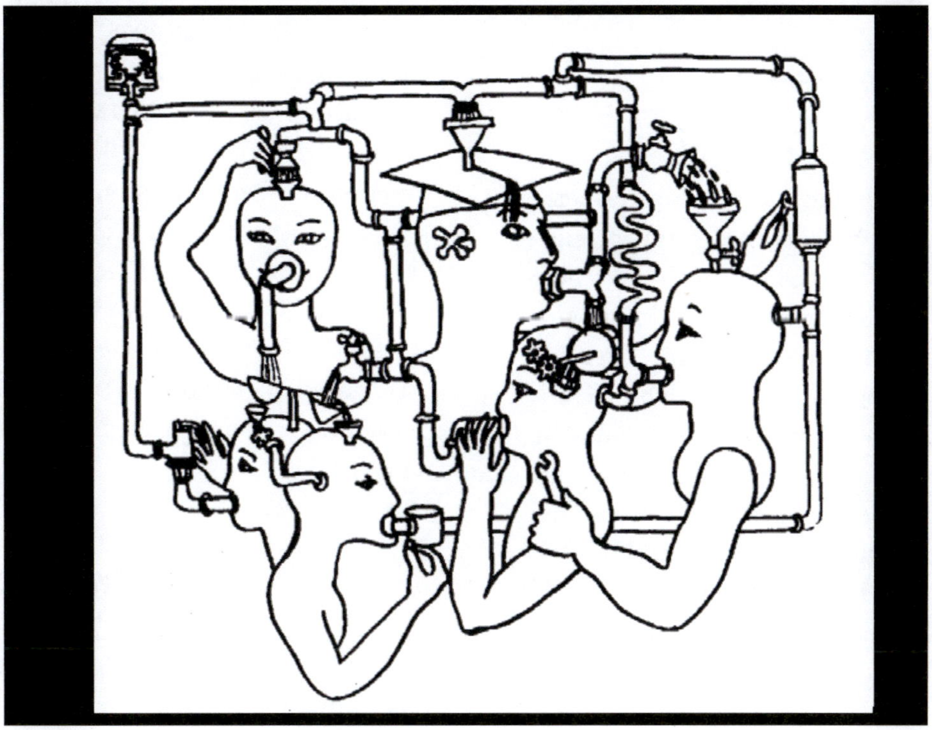

PROJECTS AND RESOURCES TO IMPROVE SCIENCE EDUCATION

BEN

BiosciEdNet (BEN) Collaborative was established in 1999 by the American Association for the Advancement of Science (AAAS) with 11 other professional societies and coalitions. The BEN Collaborative mission is not only to provide seamless access to e-resources but to also serve as a catalyst for strengthening teaching and learning in the biological sciences. BEN resources have been reviewed by the individual societies for standards of quality and accuracy; the collaborative establishment of its metadata structure permits the user to easily conduct productive interdisciplinary searches across the diverse biological sciences topics.
www.biosciednet.org

BioQuest

The BioQUEST Curriculum Consortium (BQCC) is a community of scientists, teachers, and learners who are interested in supporting biology education that reflects realistic scientific practices. The efforts in science education build on a commitment to engaging learners in a full spectrum of biological inquiry from problem posing to problem solving and peer persuasion. Many of the projects involve coordinating faculty development workshops that focus on strategies for bringing realistic scientific experiences into their classrooms and collaboratively developing curriculum projects.
http://bioquest.org/

Center for the Integration of Research, Teaching, and Learning (CIRTL)

The Center for the Integration of Research, Teaching, and Learning (CIRTL) is an NSF Center for Learning and Teaching in higher education. CIRTL uses graduate education as the leverage point to develop a national STEM faculty committed to implementing and advancing effective teaching practices for diverse student audiences as part of successful professional careers. The goal of CIRTL is to improve the STEM learning of all students at every college and university, and thereby to increase the diversity in STEM fields and the STEM literacy of the nation.

To prepare the future STEM faculty of the nation, CIRTL influences graduate-through-faculty preparation in teaching and learning at a significant number of research universities. Building on the CIRTL Core ideas, the project proposes to achieve this goal through a learning community of diverse research universities mutually engaged in teaching-as-research activities.

Established in fall 2006, the CIRTL Network was comprised of Howard University, Michigan State University, Texas A&M University, University of Colorado at Boulder, University of Wisconsin-Madison, and Vanderbilt University. After a substantial expansion in 2011, the Network now includes 25 research universities across the nation. The diversity of these institutions—private/public; large/moderate size; majority-/minority-serving; geographic location—is by design aligned with CIRTL's mission.
www.cirtl.net/

MicrobeWorld

Established in 2003, MicrobeWorld is an interactive multimedia educational outreach initiative from the American Society for Microbiology (ASM) that promotes awareness and understanding of key microbiological issues to adult and youth audiences and showcases the significance of microbes in our lives. The various

outreach methods feature the process of discovery, historical changes in research, and a variety of scientific careers in industry, academia, and government.
www.microbeworld.org

MERLOT

Multimedia Educational Resource for Learning and Online Teaching (MERLOT) is a free and open online community of resources designed primarily for faculty, staff and students of higher education from around the world to share their learning materials and pedagogy. MERLOT is a leading edge, user-centered, collection of peer-reviewed higher-education online learning materials, catalogued by registered members and a set of faculty development support services. MERLOT's strategic goal is to improve the effectiveness of teaching and learning by increasing the quantity and quality of peer-reviewed online learning materials that can be easily incorporated into faculty-designed courses.
www.merlot.org

PKAL

Project Kaleidoscope (PKAL) is one of the leading advocates in the United States for what works in building and sustaining strong undergraduate programs in the fields of science, technology, engineering, and mathematics (STEM). As an intelligence broker within the undergraduate STEM community, PKAL disseminates resources that advance the work of academic leaders tackling the challenging work of ensuring that the undergraduate STEM learning environment serves 21st century students, science, and society most effectively, efficiently, and creatively. PKAL themes include institutional transformation, human and physical infrastructure, the academic program, pedagogical tools, the national context, and twenty-first century student education.
www.pkal.org

SENCER

Science Education for New Civic Engagements and Responsibilities (SENCER) was initiated in 2001 under the National Science Foundation's CCLI national dissemination track. Since then, SENCER has established and supported an ever-growing community of faculty, students, academic leaders, and others to improve undergraduate STEM (science, technology, engineering, and mathematics) education by connecting learning to critical civic questions. SENCER's goals are to: (1) get more students interested and engaged in learning in science, technology, engineering, and mathematics (STEM) courses, (2) help students connect STEM learning to their other studies, and (3) strengthen students' understanding of science and their capacity for responsible work and citizenship.
www.senser.net

Workshop for New Physics and Astronomy Faculty

Since 1996, the American Association of Physics Teachers has sponsored workshops designed to help new faculty at research and four-year institutions understand how to become more effective educators and support their quest to gain tenure. Because of the pressure to establish their credentials in research or other scholarly activities, new faculty may be tempted to postpone or ignore the development of teaching proficiency. They may receive direct or subtle messages suggesting that only a focus on research will result in career advancement, and there is often a lack of mentors or role models who demonstrate dedication and enthusiasm for teaching.

Similar signals are transmitted to graduate students who may be in training for academic careers. Moreover, because the research universities include many of our large public institutions, a large number of undergraduates may suffer as a result of inadequate preparation of new faculty for teaching.

Data suggest that this inadequate attention to teaching, especially in introductory science and math courses, is responsible for driving students away from undergraduate majors in science, mathematics and engineering.

To improve the quality of physics teaching on a national scale, AAPT created the New Faculty Workshop. Each workshop presents a small number of techniques that have proven to be effective in a variety of environments. These tactics can be implemented with minimal time and effort, thus allowing new faculty to devote more of their attention to research and scholarship.

Each spring and fall, department chairs at research and four-year institutions are asked to nominate tenure-track faculty in the first few years of their initial appointment. The ideal candidate would have a year or two of teaching experience so that they are aware of the challenges of the first year of teaching.

In 2002, the American Physical Society and the American Astronomical Society joined with AAPT to expand the reach of this program. Financial support is provided by the National Science Foundation.

www.aapt.org/Conferences/newfaculty/nfw.cfm

REFERENCES

Bernhard, J. 2001. Does active engagement curricula give long-lived conceptual understanding? pp. 749-752. In: Physics Teacher Education Beyond 2000, R. Pinto and S. Surinach, eds. Paris: Elsevier.

Brewer, C., and D. Smith, eds. 2011. Vision and Change in Undergraduate Biology Education. Washington: American Association for the Advancement of Science.

Handelsman, J., S. Miller, and C. Pfund. 2006. Scientific Teaching. San Francisco: Freeman and Sons.

Knight, J.K., and W.B. Wood. 2005. Teaching more by lecturing less. Cell Biology Education 4:298-310.

Michael, J. 2006. Where's the evidence that active learning works? Advances in Physiology Education 30:159-167.

Meltzer, D., and R. Thornton. 2012. Resource Letter ALIP-1: Active-Learning Instruction in Physics. American Journal of Physics 80(6):478.

NRC (National Research Council). 2000. How People Learn: Brain, Mind, Experience, and School (Expanded Edition). Washington: National Academy Press.

NRC. 2003. BIO2010: Transforming Undergraduate Education for Future Research Biologists. Washington: National Academies Press.

NRC. 2004. Biotechnology Research in an Age of Terrorism. Washington: National Academies Press.

NRC. 2006b. America's Lab Report: Investigations in High School Science. Washington: National Academies Press.

NRC. 2007b. *Taking Science to School.* Washington: National Academies Press.

NRC. 2008. Ready, Set, Science! Washington: National Academies Press.

NRC. 2009c. On Being a Scientist. Washington: National Academies Press.

NRC. 2011d. Promising Practices in Undergraduate Science, Technology, Engineering, and Mathematics Education:

Summary of Two Workshops. Washington: National Academies Press.

NRC. 2011e. Research in the Life Sciences with Dual Use Potential: An International Faculty Development Project on Education about the Responsible Conduct of Science. Washington: National Academies Press.

NRC. 2012b. A Framework for K-12 Science Education: Practices, Crosscutting Concepts, and Core Ideas. Washington: National Academies Press.

Prince, M. 2004 Does active learning work? A review of the research. Journal of Engineering Education 93(3):223-231.

Springer, L., M. Stanne, S. Donovan. 1999. Measuring the Success of Small-Group Learning in College-Level SMET Teaching: A Meta-Analysis. Madison: National Institute for Science Education, University of Wisconsin. Available at *http://www.wcer.wisc.edu/archive/cl1/cl/resource/scismet.htm*; accessed June 17, 2013.

Springer, L., M. Stanne, and S. Donovan. 1999. Effects of small-group learning on undergraduates in science, mathematics, engineering, and technology: A meta-analysis. Review of Educational Research 69(1):21-51.

Appendix E
Biographies of Committee Members and Staff

Rita R. Colwell, *Chair*, University of Maryland; Canon U.S. Life Sciences, Inc.
Rita Colwell is Chairman of Canon U.S. Life Sciences, Inc. and Distinguished University Professor at both the University of Maryland at College Park and Johns Hopkins University Bloomberg School of Public Health. Her interests are focused on global infectious diseases, water, and health, and she is currently developing an international network to address emerging infectious diseases and water issues, including safe drinking water for both the developed and developing world. Dr. Colwell has held many advisory positions in the U.S. government, nonprofit science policy organizations, and private foundations, as well as in the international scientific research community. Dr. Colwell is a member of the National Academy of Sciences, the Royal Swedish Academy of Sciences, Stockholm, the American Academy of Arts and Sciences, and the American Philosophical Society. She holds a BS in bacteriology and an MS in genetics from Purdue University, and a PhD in oceanography from the University of Washington.

Enriqueta C. Bond, Burroughs Wellcome Fund
Enriqueta Bond, PhD, retired in August 2008 as President of the Burroughs Wellcome Fund (BWF), a private foundation whose mission is to advance the medical sciences through the support of research and education. She is a founding partner of QE Philanthropic Advisors and consults with philanthropic and nonprofit organizations on program development and governance. Before that she served for nearly 20 years as staff officer and division director at the Institute of Medicine, serving as executive officer from 1989 to 1994.

Dr. Bond serves on numerous board and advisory groups such as the Council of the National Institute of Allergy and Infectious Diseases, the Institute of Medicine Committee to Review the Clinical and Translational Science Awards (CTSA) Program at the National Center for Advancing Translational Science, the National Research Council Committee on Developing a Framework for an International Faculty Development Project on Education about Research in the Life Sciences with Dual Use, the Burroughs Wellcome Fund Career Award for Science and Mathematics Advisory Committee, and the Board of the Health Effects Institute and the James B. Hunt Jr. Institute for Educational Leadership.

Dr. Bond chairs a National Academies Board on Developing the Capacity of African Academies of Science, serves as a member of the Institute of Medicine Forum on Microbial Threats to Health, and is a frequent reviewer of Academy reports. She previously chaired the Institute of Medicine Clinical Research Roundtable and was a member of the Council of the Eunice Shriver National Institute of Child Health and Human Development. Dr. Bond is a member of the Institute of Medicine and is a fellow of the Association for the Advancement of Science. She was educated at Wellesley College

(AB), the University of Virginia (MA), and Georgetown University, where she earned a PhD in genetics and molecular biology.

John D. Clements, Tulane University
Dr. Clements is a professor of microbiology and immunology at Tulane University School of Medicine and director of the Tulane Center for Infectious Diseases. After receiving his doctorate in 1979 from the University of Texas Health Science Center at Dallas, he completed a National Research Council Associateship at Walter Reed Army Institute of Research in Washington, D.C. In 1980, Dr. Clements was appointed an assistant professor in the Departments of Microbiology and Medicine at the University of Rochester School of Medicine in New York. In 1982, he joined the faculty at Tulane University, where he has served as professor and chair of the Department of Microbiology and Immunology since 1999. He was Vice Dean for Research from 2006 to 2009 and in 2009 was appointed director of the Tulane Center for Infectious Diseases. Dr. Clements maintains an active research program focused on development of vaccines against infectious diseases. His research has been continuously funded from a variety of Public Health Service and Department of Defense sources. He is currently Director of the Tulane/Xavier Vaccine Development/ Engineering Project and the Tulane/Xavier Vaccine Peptide Program, both supported by the Department of Defense. Dr. Clements is also Co-Director of the South Louisiana Institute for Infectious Disease Research and Co-Director of the Louisiana Vaccine Center, both collaborative projects between Tulane University and Louisiana State University Health sciences Center in New Orleans. Research in Dr. Clements's laboratory has resulted in more than 100 peer-reviewed publications and book chapters, and 13 patents. Dr. Clements has served on numerous scientific panels and editorial boards. He currently serves on the scientific advisory boards of the Western Regional Center for Excellence in Biodefense Research and the PATH Enteric Vaccine initiative. In 2003, he was trained as a U.N. Weapons Inspector (Biologic) in the 7th United Nations Monitoring, Verification and Inspection Commission (UNMOVIC). In 2003 and again in 2004, he served as a member of the Iraq Survey Group as a subject matter expert in weapons of mass destruction and dual use equipment and programs.

Nancy D. Connell, University of Medicine and Dentistry of New Jersey
Dr. Connell is a professor in the Division of Infectious Disease in the Department of Medicine at the University of Medicine and Dentistry of New Jersey (UMDNJ), New Jersey Medical School. A Harvard University PhD in Microbiology, Dr. Connell's major research focus is the interaction between respiratory infectious agents and the macrophage. She is director of the Biosafety Level Three (BSL-3) Facility of UMDNJ's Center for the Study of Emerging and Re-emerging Pathogens and chairs the University's Institutional Biosafety Committee. She has served on a number of National Academies committees, e.g., the Committee on Advances in Technology and the Prevention of their Application to Next Generation Biowarfare Agents and the Committee to Review the Scientific Approaches used in the FBI's Investigation of the 2001 *Bacillus anthracis* Mailings.

Clarissa Dirks, The Evergreen State College
Clarissa Dirks is an associate professor in scientific inquiry, biology at the Evergreen State College in Olympia, Washington. She earned her PhD in molecular and cellular biology at the University of Washington, conducting research in virology at the Fred Hutchinson Cancer

Research Center. As a virologist she currently investigates the evolution of viruses and host viral inhibitory proteins as well as the coevolution of bryophytes and species of Tardigrada. As a biology education researcher, she has implemented programs to improve retention of underrepresented students in first-year science courses, and conducted studies to better understand how students acquire and master science process skills. She has received two Tom Rye Harvill Awards for the Integration of Art and Science, has been named a National Academies Education Fellow and Mentor in the Life Sciences, and is the recipient of two Biology Leadership Education grants. She works to provide professional development opportunities for faculty and postdoctoral scholars by serving on the Committee for National Academies Summer Institute on Undergraduate Education in Biology, leading a Pacific Northwest Regional Summer Institute, and mentoring postdoctoral fellows as a regional field station leader for the Faculty Institute for Reforming Science Teaching. She is a member of the editorial board of the journal *CBE-Life Science Education* and a cofounder of the Society for Biology Education Research (SABER).

Mohamed El-Faham, Bibliotheca Alexandrina
Mohamed El-Faham is director of the Center for Special Studies and Programs (CSSP), Bibliotheca Alexandrina, Egypt. He is also a professor and director of Power Systems Group at the Department of Electrical and Computer Control Engineering, Faculty of Engineering and Technology, Arab Academy for Science and Technology and Maritime Transport in Alexandria. He received his BSc in electrical engineering from the University of Alexandria and his MSc and DSc in electrical engineering from the George Washington University, Washington, D.C. He is a senior member of the Institute of Electrical and Electronics Engineers (IEEE) and the author/coauthor of a number of publications. As director of the CSSP, he organizes, each year, a number of major conferences in the fields of science, technology, and education.

Alastair Hay, University of Leeds
Alastair Hay is professor of environmental toxicology in the School of Medicine at the University of Leeds, U.K. He holds a BSc in chemistry and PhD in biochemistry, both from the University of London. As a toxicologist his major interests are the effects of chemicals on health but his research also covers work on calcium metabolism, kidney damage, drugs of abuse, pharmacokinetics, and proteomics. Professor Hay currently teaches basic biology, research methodology, and ethics to medical students in years 1 to 3 of their 5-year medical degree. External to the university he has been an advisor to the U.K. government for over 20 years on both the regulation of chemicals and exposure standards in the workplace; he also advises the European Union on workplace exposure limits. He has more than 35 years' experience with chemical weapons issues and advises the U.K. government on matters relating to the implementation of the 1997 Chemical Weapons Convention. He has developed teaching materials for chemists on such topics as multiple uses of chemicals; chemical weapons; and codes of conduct. Professor Hay has worked with numerous national and international organizations to promote these issues in both the chemical and biological sciences and to help find innovative teaching approaches to engage young scientists and promote responsible conduct in research.

Elizabeth Heitman, Vanderbilt University Medical Center
Dr. Heitman received her PhD from Rice University in 1988. She has extensive expertise in

biomedical ethics, responsible conduct of research, and ethics in public health, as well as experience with biodefense-related ethical decision making as a member of the Policy, Ethics, and Law Core of the Southeast Regional Center of Excellence for Emerging Infections and Biodefense (SERCEB). Her primary research addresses the evaluation of education in the responsible conduct of research, and the cultural awareness and professional socialization of students and researchers. Dr. Heitman is the director of a four-year research ethics education program for Costa Rican biomedical researchers and research ethics review committees sponsored by the NIH's Fogarty International Center and a member of the Clinical Research Ethics Key Function Committee of the Clinical and Translational Science Award (CTSA) Consortium. She is the coauthor of *The Ethical Dimensions of the Biological and Health Sciences* (with Drs. Ruth Ellen Bulger and Stanley Joel Reiser).

Adel A.F. Mahmoud, Princeton University
Adel A.F. Mahmoud, MD, PhD, is a professor in molecular biology and public policy at Princeton University, and former president of Merck Vaccines of Merck & Company, Inc. Before that, he served at Case Western Reserve University and University Hospitals as chairman of medicine and physician in chief. Dr. Mahmoud's academic pursuits focused on investigations of the determinants of infection and disease in human schistosomiasis and helminthic infections. He has led efforts to develop new vaccines for measles, mumps, rubella, varicella, rotavirus, shingles, and human papillomavirus. Dr. Mahmoud served as a member of the National Science Advisory Board for Biosecurity. He was elected to the Institute of Medicine (IOM) of the National Academy of Sciences in 1987 and has served on numerous committees. For example, his leadership in setting global health strategies shaped the agenda of the IOM Forum on Microbial Threats by tackling such topical issues as biological threats and bioterrorism; SARS; and pandemic flu. He received an MD from the University of Cairo and a PhD from the University of London, School of Hygiene and Tropical Medicine.

Mona Mostafa Mohamed, Cairo University
Mona Mostafa Mohamed, PhD, is professor of cell biology and head of the Cancer Biology Research Laboratory, Faculty of Science, Cairo University. Upon completion of her doctorate at Cairo University, she was competitively selected for a prestigious Avon-AACR International Scholar award in breast cancer research (2005-2007), one of only 12 selected from several hundred applicants. Dr. Mohamed's research focuses on the interactions between inflammatory macrophages and their associated cytokines and proteolytic enzymes observed during breast cancer, with the ultimate goal of understanding mechanisms by which macrophages induce breast cancer progression and identifying novel targets for drug development. Returning to Egypt in 2007, she was awarded start-up funds from Avon Foundation and Cairo University to establish the first specified breast cancer biology laboratory in Egypt (CBRL; *www.cbrl.cu.edu.eg*). CBRL's state-of-the-art equipment has enabled Dr. Mohamed's group to achieve outstanding results in breast cancer research, including those of 13 master's and doctoral students. Dr. Mohamed is the recipient of numerous grants from the Science and Technology Development Fund, Egypt; the Avon Foundation (U.S.A.) in collaboration with New York University; the Fogarty International Research Collaboration - Basic Biomedical (FIRCA-BB) Research Award (R03); and Wayne State University (U.S.A.). Dr. Mohamed is a leading example for women in science, blazing a path forward for future

women seeking scientific and academic careers. She was recently selected for the 2012 Women in Science Hall of Fame for her scientific accomplishments (*http://jordan.usembassy.gov/wshf_2012.html*).

James H. Stith, American Institute of Physics
James H. Stith is Vice President Emeritus for the American Institute of Physics (AIP). While an officer of the Institute, he had oversight responsibilities for AIP's Magazine Division, the Media and Government Relations Division, the Education Division, the Center for the History of Physics, the Statistical Research Division, and the Careers Division. Throughout his career, he has been an advocate for programs that ensure ethnic and gender diversity in the sciences. His doctorate in physics was earned from the Pennsylvania State University, and his master's and bachelor's in physics were received from Virginia State University. A physics education researcher, his primary interests are in program evaluation, and teacher preparation and enhancement. He was formerly a professor of physics at the Ohio State University and professor of physics at the United States Military Academy. A retired colonel, he was the first African American to earn tenure at the Academy. Dr. Stith has been a visiting associate professor at the United Air Force Academy, a visiting scientist at the Lawrence Livermore National Laboratory, a visiting scientist at the University of Washington, and an associate engineer at the Radio Corperation of America. He is a past president of the American Association of Physics Teachers, past president of the National Society of Black Physicists, a fellow of the American Association for the Advancement of Science, a fellow of the American Physical Society, a chartered fellow of the National Society of Black Physicists, and a member of the Ohio Academy of Science. He was named a distinguished alumnus of the Pennsylvania State University (the Alumni Association's highest award); an honorary member of Sigma Pi Sigma (its highest award), the physics honor society; and a National Academies Education Mentor in the Life Sciences. He was recognized by Science Spectrum Magazine as one of the 50 Most Important Blacks in Research Science and was named a ScienceMaker by HistoryMakers. Additionally, he serves on a number of national and international advisory boards and has been awarded a Doctor of Humane Letters by his alma mater, Virginia State University. He is married and has three adult daughters and two grandchildren.

National Academies Staff

Lida Anestidou is senior program officer at the Institute for Laboratory Animal Research of the U.S. National Academy of Sciences, where she directs a diverse portfolio of studies on the use of laboratory animals; biodefense and biosecurity; and research integrity/responsible conduct of research. Prior to this position she was faculty at the Center for Biomedical Ethics and Society, Vanderbilt University Medical Center. She earned her doctorate in biomedical sciences from the University of Texas at Houston. Working with physiologist Norman Weisbrodt, she explored the effects of nitric oxide on the motility of the gastrointestinal musculature. Working with research integrity expert and biomedical ethics educator Elizabeth Heitman, she concurrently pursued her interests in biomedical ethics, scientific integrity, and science policy. Dr. Anestidou also holds a Doctor of Veterinary Medicine degree from Greece (her home country) and an MS in Veterinary Sciences from the University of Florida. She is an editorial board member of Science and Engineering Ethics, Lab Animal, and SciTech Lawyer and an ad hoc reviewer for the American Journal of Bioethics. She is a member of the National Conference of Lawyers and Scientists. Dr. Anestidou serves as an expert reviewer in the Ethics Evaluation of grant applications to the 7th Framework Program of the European Research Council and the European Commission Directorate General Research.

Jo L. Husbands is a scholar/senior project director with the Board on Life Sciences of the U.S. National Academy of Sciences (NAS), where she manages studies and projects to help mitigate the risks of the misuse of scientific research for biological weapons or bioterrorism. She represents the NAS on the Biosecurity Working Group of IAP: The Global Network of Science Academies, which also includes the academies of Australia, China, Cuba, Egypt, India, Nigeria, Poland (chair), Russia, and the United Kingdom. From 1991 to 2005 she was director of the NAS Committee on International Security and Arms Control (CISAC) and its Working Group on Biological Weapons Control. Before joining the National Academies, she worked for several Washington, D.C.-based nongovernmental organizations focused on international security. Dr. Husbands is currently an adjunct professor in the Security Studies Program at Georgetown University. She is a member of the Temporary Working Group on Education and Outreach in Science and Technology of the Organization for the Prohibition of Chemical Weapons and the Global Agenda Council on Nuclear, Chemical, and Biological Weapons of the World Economic Forum. She is also a fellow of the International Union of Pure and Applied Chemistry. She holds a PhD in political science from the University of Minnesota and a master's in international public policy (international economics) from the Johns Hopkins University School of Advanced International Studies.

Jay B. Labov is senior staff member of the National Research Council's Center for Education. In this capacity, he leads an institution-wide effort to leverage the National Academies' work in education by helping to make more deliberate connections between the work of the Center for Education, the National Academy of Sciences, National Academy of Engineering, and the program units of the National Research Council. He is the principal liaison on education activities between the program units of the National Academies and its Office of Communications, with the goal of enhancing communication with outside stakeholders about the Academies' work in education and the public's understanding of science and technology. He also has been the study director for several NRC reports: Evaluating and Improving Undergraduate Teaching in Science, Mathematics, Engineering, and Technology (2003); Learning and Understanding: Improving Advanced Study of Mathematics and Science in U.S. High Schools (2002); Educating Teachers of Science, Mathematics, and Technology: New Practices for the New Millennium (2000); Transforming Undergraduate Education in Science, Mathematics, Engineering, and Technology (1999); Serving the Needs of Pre-College Science and Mathematics Education: Impact of a Digital National Library on Teacher Education and Practice (1999); and Developing a Digital National Library for Undergraduate Science, Mathematics, Engineering, and Technology Education (1998). He has been Director of the Center's Committee on Undergraduate Science Education and oversees the National Academy of Science's efforts to improve the teaching of evolution in the public schools. Prior to assuming his position at the NRC Dr. Labov was a member of the biology faculty for 18 years at Colby College in Waterville, Maine.

Appendix F

Educational Institute for Responsible Research on Infectious Diseases: Ensuring Safe Science in the 21st Century

Aqaba, Jordan, September 7-13, 2012
Participants

ALGERIA

Halima Benbouza, Head, National Biotechnology Research Center, Constantine

Abdelkader Bouyakoub, Professor, Department of Mathematics, Faculty of Sciences, University of Oran (Es-Sénia), Oran

Ben Amar Cheba, Lecturer, Department of Biotechnology, Faculty of Sciences, University of Science and Technology-Oran-Mohamed Boudiaf, Oran

Noureddine Yassaa, Professor, Faculty of Chemistry, University of Sciences and Technology, Houari Boumediene, Algiers

EGYPT

Amal Abd EL Raof,Chief Researcher,Virology Department, Animal Health Research Institute, Cairo

Mahmmoud Sayed Abd El-sadek, Lecturer, Physics Department, Faculty of Science, South Valley University, Qena

Marwa Ahmed Ali Abd EL Wahab, Lecturer, Faculty of Medicine, Tanta University, Tanta

Yahya Al-Naggar, Assistant Lecturer, Zoology Department, Faculty of Science, University of Tanta, Tanta

Manal Eid, Associate Professor, Department of Botany, Faculty of Agriculture, Suez Canal University, Ismailia

Yahya Zakaria Eid, Associate Professor, Department of Poultry Production, Faculty of Agriculture, Kafrelsheikh University, Kafr El-Sheikh

Mohamed Mostafa Ahmed Elhadidy, Lecturer, Faculty of Veterinary Medicine, Mansoura University, Mansoura

Mohamed Ahmed Ellabban, Lecturer, Faculty of Medicine, Suez Canal University, Ismailia

Mohamed El-Sayed El-Shinawi, Associate Professor, Faculty of Medicine, Ain Shams University, Cairo

Mohammed Salah El-Tholoth, Lecturer, Faculty of Veterinary Medicine, Mansoura University, Mansoura

Mohammad Mustafa M. Ibrahim, Associate Professor, College of Biotechnology, Misr University for Science and Technology, Giza

Fatma Salem, Botany Department, Faculty of Science, Suez Canal University, Ismailia

Mohamed Labib Salem, Professor, Zoology Department, Faculty of Science, Tanta University, Tanta

Yaldez Zein ElDin, Lecturer, Faculty of Nursing, Damanhour University, Alexandria

JORDAN

Amjed Al-Fahoum, Electronic Engineering Department, King Abdullah II School for Engineering, Princess Sumaya University for Technology, Amman

Saied Jaradat, Director, Princess Haya Biotechnology Center, Jordan University of Science and Technology, Irbid

Khalid M. Al-Batayneh, Associate Professor, Department of Biological Sciences, Yarmouk University, Irbid

LIBYA

Abubaker Toboli, Head, Department of Microbiology and Parasitology, Faculty of Medicine, University of Benghazi, Benghazi

YEMEN

Samira Al-Eryani , Assistant Professor, Department of Medical Parasitology, Faculty of Medicine and Health Sciences, Sana'a University, Sana'a

Khaled Abdulla Al-Sakkaf, Assistant Professor, Department of Community Medicine and Public Health, Faculty of Medicine and Health Sciences, Aden University, Aden

Huda Omer Ba Saleem, Assistant Professor, Department of Community Medicine and Public Health, Faculty of Medicine and Health Sciences, Aden University, Aden

Amen Bawazir, Associate Professor, College of Medicine, Khormaksar, University of Aden, Aden

Ahmed Moharem, Assistant Professor, Department of Medical Laboratory, College of Medicine and Health Sciences, Thamar University

Qais Abdullah Nogaim, Assistant Professor, Department of Food Science and Technology, Faculty of Agriculture, Ibb University

APPENDIX G

Documents from the Institute

Call for Applications

Educational Institute on Responsible Research with Infectious Diseases:
Ensuring Safe Science in the 21st Century
8-13 September 2012, Aqaba, Jordan

Bibliotheca Alexandrina in cooperation with the National Research Council of the U.S. National Academy of Sciences announces the call for applications for the ***Educational Institute on Responsible Research with Infectious Diseases: Ensuring Safe Science in the 21st Century***.

The Institute will take place in Aqaba, Jordan from 8-13 September 2012. It aims to develop a network of faculty in the Middle East–North Africa (MENA) region able to teach issues related to research with dual use potential by using tenets of responsible science and active learning pedagogical techniques.

Eligibility Criteria

- Applicants must be faculty who have been working and living in **Egypt, Yemen, Algeria or Abu Dhabi and should apply in teams of 2-3 from each institution.**
- Applicants must stay for the entire institute from 8-13 September 2012.
- Applicants **must** agree to use one or more of the instructional materials developed at the Institute in their teaching, preferably in the fall semester 2012.
- It is expected that one member/team will participate in the follow-up evaluation meeting during the academic year 2012-2013.

The following documents MUST be submitted:

1. Application form (click here)
2. Personal statement (no more than one page) that includes:
    a. Your interest in scientific teaching
    b. The types of courses you teach
    c. What you hope to achieve by attending the Institute

Only complete applications will be considered
Deadline: July 23, 2012

Post-Institute Survey

As a participant in the recent *Educational Institute on Responsible Research with Infectious Diseases: Ensuring Safe Science in the 21st Century* that was held in Aqaba, Jordan, from 7-13 September, 2012, your reflections on the quality of discussions during the Institute and the implications of these discussions for research, policy, and education practice are important.

We are inviting you to provide feedback to the U.S. National Academy of Sciences and to the Bibliotheca Alexandrina about the Institute itself. We also would like to know how you are planning to use what you learned at the Institute in your own academic/research setting and in collaboration with other participants from your country and across the region. While your response will remain confidential, your views will be combined with those of your Institute colleagues to guide us in improving future programs on education about responsible conduct of science. This survey should take approximately 15 minutes to complete.

Thank you for your participation.

Lida Anestidou and Jay Labov, on behalf of the organizing committee of the U.S. National Academy of Sciences

Mohamed M. El-Faham, Director, Center for Special Studies & Programs, Bibliotheca Alexandrina

---

1. Which ONE of the following best describes your position in your institution?

    ___Faculty or Lecturer for Undergraduate Students
    ___Faculty or Lecturer for Graduate or Postdoctoral Students
    ___Academic Administrator
    ___Other (Please specify): ____________________________________

2. Please indicate which THREE reasons best describe why you chose to attend this Institute:

    ___To meet colleagues from my country who share interests in responsible conduct of science
    ___To meet colleagues from other countries who share interests in responsible conduct of science
    ___To reconnect with colleagues who share my interest in responsible conduct of science
    ___To deepen my understanding of the issues related to the responsible conduct of science
    ___To discover tools, resources and best practices for incorporating evidence-based teaching techniques into my courses
    ___To become more involved with future efforts to improve education about the responsible conduct of research in my country
    ___To become more involved with future efforts to improve education about the responsible conduct of research internationally
    ___Other (please specify): ____________________________________

3. This Institute was designed to bring people together from across the Middle East/North Africa region and hopefully to launch a series of future activities to promote education about the responsible conduct of

science. Based on discussions at the Institute and recommendations generated by participants, please indicate which THREE of the following next steps you consider as the most important priorities for maintaining momentum and moving forward to infuse education about the responsible conduct of science into courses and other educational programs that are taught by you and your colleagues at your institution?

___ Provide at least one similar workshop for current faculty colleagues at your institution
___ Provide at least one similar workshop for future faculty colleagues, i.e., graduate and post-graduate students at your institution
___ Engage the leadership at your institution to enable you to create a permanent curricular modification that allows you to provide such workshops on a regular basis
___ Create a network of faculty from your institution and from at least one more institution that will be in charge of coordinating and teaching similar workshops
___ Lay the groundwork for proposals to seek support from various funding agencies for you and your institution to support education about the responsible conduct of science.
___ Other: ______________________________

4. How are you planning to move forward on the issues addressed at the Institute?

	Am Already Using/Doing	Plan to Use/Do	Do Not Plan to Use/Do	Not Applicable
Obtain and share reports and other resources on education and issues related to the responsible conduct of science from the National Academies and the Bibliotheca Alexandrina.[65]				
Use reports and other resources from the National Academies and the Bibliotheca Alexandrina to inform your own projects on the responsible conduct of science.				
Use reports and other resources from the National Academies and the Bibliotheca Alexandrina to improve your pedagogy on the responsible conduct of science and other subject areas.				

[65] All reports from the National Academies are available for free download as pdf files at *http://nap.edu.*
Reports and other resources from the Bibliotheca Alexandrina are available at *www.bibalex.org/cssp/publications/publications.htm*
and
www.bibalex.org/Publications/BA_Publications_EN.aspx

Continue to discuss the messages and outcomes of the Institute with others in your projects or organization				
Discuss the Institute with professional colleagues outside of your organization.				
Continue to interact with Institute presenters and facilitators.				
Continue to interact with other Institute participants				
Work with professional societies and other organizations (to which I belong to encourage colleagues to employ active learning in teaching about the responsible conduct of science and related topics.				
Other (please describe):				

5. How well did the Institute meet your expectations?

Exceeded My Expectations	Met All of My Expectations	Met Some of My Expectations	Did Not Meet My Expectations

Please explain what you found particularly effective OR ineffective.

____

____

____

6. Please rate the Institute on:

	Excellent	Very Good	Good	Fair	Poor
Value of the Institute as a learning or professional development experience					
Clarity of Institute's goals and					

objectives					
Relevance of topics that were presented in relation to the stated goals of the Institute					
Quality of sessions about the responsible conduct of science					
Quality of sessions about the scientific basis for the use of active learning techniques					
Relevance to you and your work of the issues presented					
Usefulness of resources provided by the organizers and presenters (e.g., background resources in the Dropbox and briefing book)					
Inclusion of information and perspectives from a diverse range of views					
Balance of time spent in whole group and team breakout sessions					
Amount of time devoted to discussions during plenary sessions					
Time to meet and interact with other participants					
Helpfulness of your breakout group's facilitators					
Helpfulness of the National Academies staff					
Hotel accommodations and meals					

7. If the National Academies were to organize and host additional Institutes or related activities on this topic in the future, would you be interested in participating?

____Definitely yes _______ Maybe _______ No

If yes, what role would you see for yourself in such activities?

8. Please offer suggestions about what to incorporate or avoid in future National Academies activities related to education about the responsible conduct of science.

Your name (optional) ____

Funding Available to Help Implement Teaching about the Responsible Conduct of Science

Funding up to $1500 to help implement teaching about the responsible conduct of science (RCS) is available to participants who attended the recent Institute in Aqaba, Jordan, organized by the U.S. National Academy of Sciences and the Bibliotheca Alexandrina. Those wishing to be considered for an award should submit an application of **no more than 2 pages** explaining how they propose to implement the teaching of RCS.

Applications need to describe the following:

Applicant(s) - Applicant(s) or team responsible for conducting, managing and coordinating the project
Goal(s) - Describe the overall learning goals for what you wish to implement
Objectives - List the objectives of the RCS teaching event and briefly describe how you will assess if you were successful in meeting those objectives
Approaches—indicate the teaching methods to be used
Participants—Describe the expected audience (i.e., colleagues, postdoctoral fellows or postgraduates) and the reason it was chosen
Budget - This should detail anticipated costs and any funding/support which will be available from your institution to offset costs. Explain what resources your institution is willing to provide (space, support, etc.)
Timeline—Describe the expected timeframe in which you will conduct and assess the success of your project
Anticipated Problems—Identify any anticipated difficulties and comment on how these difficulties might/will be addressed
Sustainability - Explain how you will attempt to sustain teaching about RCS at your institution and, where possible, promote it in your country of residence.

Funds will be allocated as a U.S. National Academy of Sciences award to individuals and will provide **no support for institutional costs**. Although the principal applicant must have been a participant at the Aqaba Institute, the application may name other individuals who will work with you on the project but who did not attend the Institute. Joint applications from several participants at the Aqaba workshop who propose to work together will be particularly welcome.

The deadline for receipt of applications is **OCTOBER 18, 2012.**

The proposals will be reviewed by the project's committee members and individuals who will receive awards will be notified on or about November 1, 2012.